MORSE CODE
how to learn and make radio contacts
cw4u.org

David González EA7HYD
- Nov. 2019 -

www.cw4u.org
cw4u@hotmail.com

© 2019 David González, EA7HYD.

ISBN: 9781698906508
Imprint: Independent publication.

Text, illustrations, design, layout, cover:
David González, EA7HYD.

Original in Spanish, translation of: Jesús Pacheco González.
Corrections: Michael Sansom, G0POT.

2017 November. 1st version printed in Spanish.
2019 November. Printing and distribution from Amazon.

All the content from this document have a mere informational value; author's ideas or gathered from different sources with the aim of disseminating the general topic. Images are author's property, but those obtained from internet are properly quoted in case authorship is known. The author declines any claim for damages relative to the provided information from this document and the attached files.

INDEX

INTRODUCTION 9

*THE HISTORY
OF DOTS AND DASHES* 10
Morse code is still in use

WHAT DO WE NEED 12
to learn

HOW TO LEARN THE CODE 14
tools and methods

TIPS AND TRICKS 18
to study and learn the code

    about the lessons 19
    some advice 20

CW OR MORSE CODE 23
are they the same?

RECOGNIZE 24
the sounds, space and timing

WORDS PER MINUTE 25
accuracy transcends speed

ETIQUETTE 27
rules, codes and patterns

    Q codes
    many things with only three letters . 28
    Abbreviations
    the economy of language 29
    Prosigns
    procedural signs 30
    RST, signal report
    reporting radio signal quality 32
    Information order
    QSOs do have a structure 33

MAKING A QSO 35
real contact in Morse code

    Real contacts in Morse code 36
    Analysing QSOs
    what could have been done 38
    starting a QSO 39
    they haven't heard our callsign 41
    answering a CQ 42
    contact established, sending data .. 43
    we continue and answer 45
    farewell and closing 46
    extending the QSO 50
    The basics when on-air
    newbie, not a dummie 52
    <BK> quick exchange and its use
    usage and instances 54

TRANSMIT 56
writing with good "handwriting"

positioning and handling the key 57
straight key 58
single, sideswipers, twin paddles .. 59
iambic paddle taps 60
mechanical bug 61

Starting to "tap"
sending is different to receiving 62
connecting the key 63
dots and dashes generator 63
practice with an oscillator 64
monitor and correct your code 64
key maintenance 64

PROCEDURES 66
to keep in mind

Specific CQ calls
searching for a particular contact ... 67

Contacts in a contest
organized chaos 68

Working Split mode, Pile-ups
a QSO that it is not what it appears .. 70

Miscellany
if in doubt, ask 72

ANNEX 75
curiosities, information

Morse code table 76
Your first QSO template 77
Curiosities, history 78
Extended abbreviations 82
Extended Q codes 83
QSO examples 84
LCWO – configuration 87
SOTA 88
IARU Region1, 2. HF bands 90
Links, videos, information 92
Bibliography 93
Logbook, example page 94

*To all those people,
that will never let Morse code die.*

INTRODUCTION

Times change, even for dots and dashes. After being used for a century as a worldwide messenger service, today Morse code has been reduced to radio enthusiasts.

Until very recently, it was necessary to be able to operate with this code to obtain the radio enthusiast Ham Radio License. The most common way to learn it was through a radio elmer willing to teach it, but also from courses recorded on cassettes or discs.

Today it is not mandetory and, right after a drastic drop in the number of radio operators, Morse code has reappeared with countless useful tools with which to learn it: computer programs, web pages, mobile apps. These enable you to start recognising characters, even at a reasonable speed.

However this does not mean that we are completely ready to start making contacts on the radio. It is not only necessary to recognize letters, abbreviations or codes, but also to know the basic structures of a QSO, being able to operate with its different parts, formalities and etiquette, participating in contests, etc.

This book gathers all the information I wish I could have had when I started with CW. I have tried to make it simple and easy-to-read. It could be more extensive, but if you are interested in the topic, you have more detailed information on the internet. I hope it helps you to learn about radiotelegraphy and enjoy what even today is considered to be a unique and surprising way of radio broadcasting.

Thanks to my CW master Mikel EA2CW. This work would not have been possible without his support.

73 from David, EA7HYD

My intention is to keep this book alive thanks to your comments and contributions. Do not hesitate to send them to me via email cw4u@hotmail.com

THE HISTORY OF DOTS AND DASHES
Morse code is still in use

Morse code was invented by **Alfred Vail** in 1838 to be used on an electric telegraph in which he was collaborating with **Samuel F.B. Morse** with the aim of sending electric pulses through wires. At both ends of the wire, electromagnets were closed with the electric current **-CLICK-** marking with a dash or dot on a strip of paper depending on the duration of the closure. This strip of paper was afterwards read and decoded.

Time passed, and it became possible to identify the length of those closures, the spaces and silence between them just by listening.

This way they were able to decode the characters articulating them as "DIT" (dots) and "DAH" (dashes).

Letters were written down making words and sentences from the message. The strip of paper was no longer necessary and it started to be learned by its sound.

Initially, Morse code was meant to transmit groups of numbers that were codified in a complex "dictionary of words".

Alfred Vail not only developed the idea of dots and dashes, but also created the telegraph key.

After the patent of Morse's invention, Vail's recognition vanished.

The international Morse code was standardised in 1865 in the International Telegraph Conference in Paris. Up until that moment, different codes were used, for instance the Railway code and the American Morse code.

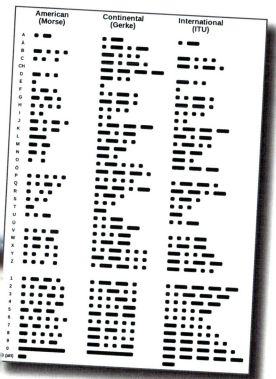

In the last decade of the 19th century **the sending of messages via cable became commonplace** (cablegram, telegram).

Some years later, radio was established as the main tool. From 1920, huge radio stations were built in zeppelins, planes and boats that oriented themselves thanks to radio beacons and parking lights. Pilots needed to know Morse code, although these radio beacons broadcasted at a very slow speed (5 wpm).

From that moment, the Dots and Dashes were used until the end of the 20th century.

Wireless telegraphy was a revolution, creating a worldwide net of commercial and governmental radiotelegraphy stations. It was mandatory in merchant ships.

This created a new and modern profession exploring the entire world, the radiotelegraph operator.

The invention also caught the attention of a few mad men interested in construction and experimentation. A group of radio hams all over the world assembled their devices and tested their aerials. They enjoyed keeping in contact with each other via Morse code.

Today, the code is still in use even though it is not mandatory in some countries to obtain the operator license. Fortunately, the code is still used by radio amateurs from around the world.

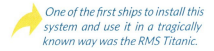
One of the first ships to install this system and use it in a tragically known way was the RMS Titanic.

WHAT DO WE NEED to learn

MOTIVATION is more important than sounds and devices.

Do not think that you are going to be the odd one who cannot learn it, many people before you have achieved it, some of them with an innate ability and others based on **PERSEVERANCE AND DEDICATION**.

Age doesn't matter, you only need to be motivated and patient during a period where you will spend 20 to 30 minutes per day practicing, no more than that. **Make it entertaining and enjoyable. After all, it's just a hobby.** You will have plenty of time to get faster afterwards and to learn to read and write the code fluently.

Don't worry if at the beginning things don't go your way, don't give up and gradually you will see improvements and realize just how much you have learned. Not everyone is great from the start... there are those that are born to play the piano while others have to work hard at it. **The end is what matters**. Some people need more time to get the same results. If it is your second attempt to learn, forget the first time and start again, confident that this time you will be successful.

CW is fun, and when you start to make contacts on the radio you will feel a great satisfaction. It is not a sprint but a long-distance race and you have to reach the finish line with determination and a steady pace.

> THE MOST IMPORTANT THINGS IS NOT TO GIVE UP AND TO DEDICATE 20 OR 30 MINUTES PER DAY FOR A FEW WEEKS.

METHOD
It has been proven that the most effective **method is KOCH**. Further information will be provided later.

TIME AND PLACE
You need a suitable and comfortable place. If not, you will have to improvise: on public transport, waiting room, on a break... wherever you want.

TIME AND CALENDAR
For some weeks you must dedicate 30 minutes to listening to Morse code lessons, every day.
It is not necessary to do more than that. Take advantage of the time you use.

ALONE OR IN A GROUP
With an APP, a computer programme, internet, MP3 files... you can study alone, but it is always better if you can do it in a group. If you are not tenacious, this will be the most suitable way of being punctual and making an effort along with your mates. It is a good option to let one of your mates be the teacher.

MOBILE PHONES, COMPUTER, MP3
Whatever you have and makes you feel comfortable.
With any of these devices you can practice with an application, program or recordings.

HEADPHONES
Headphones let you isolate yourself from the outside noise and focus. Also, you won't look like a weirdo if you are surrounded by people. The "button" like earphones can be carried in the pocket and you can practice with a mobile app. They are truly useful.

PAPER AND PENCIL, KEYBOARD
If you add a small notebook and pencil to your headphones, you already have all the necessary equipment to practice with your phone anywhere. Never erase your mistakes, because you can learn from them. Everything will be worth it in the future if you practice on your computer and use the keyboard.

KEY / OSCILLATOR - BUZZER
When you know the characters and before going on the radio, you will have to practice sending. For this you will need a "beep" generator and a key with which to make the dots and dashes. This topic has a chapter of its own.

HOW TO LEARN THE CODE
tools and methods

- **ONLINE/INTERNET APPLICATIONS**

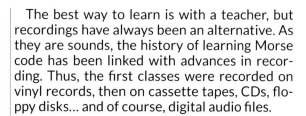

LCWO - Score: 9/10

www.lcwo.net

The easiest and most educational:

- Uses the Koch method.
- You can practice the method with lessons or text groups.
- You can check the results automatically.
- You can use charts and statistics to check your progress.
- Indicative tests, word and text groups and QTC are available.
- Turn your texts into MP3 files to record them and practice.
- It is possible to use a mobile phone browser (HTML 5).

It is easy to configure, but if you find any problem, search some help videos on YouTube or check the appendix.

The best way to learn is with a teacher, but recordings have always been an alternative. As they are sounds, the history of learning Morse code has been linked with advances in recording. Thus, the first classes were recorded on vinyl records, then on cassette tapes, CDs, floppy disks... and of course, digital audio files.

The digital archives have given rise to countless online courses and programs for computer or mobile phone, which, **being generated "mechanically", offer us an impeccable and random pattern to listen to from which we can learn perfect code.**

Applications or programs can often be free. Here I show you some of the more popular at the time of publishing. They can act as your teacher **so, from the start, whenever you have a moment, you will have a "tutor" who will help you study** (your phone and headphones will allow you to learn almost anywhere).

It will always be more motivational for you if you have a group of friends to learn with in person, on the radio or via the Internet. I have been involved in several Morse code courses held using Skype. These sessions are a simple an easy way to force yourself to reserve a specific time and day to practice, to discuss problems and if possible listen to the teacher's comments without having to depend on radio propagation.

It is not as difficult as you think and once you have learned Morse code you will be able to enjoy a part of your station that that was previously unused. You will only have to fulfill two points, which I will now comment on.

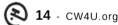

Today, almost all programs use the **KOCH METHOD** because it is the most recognized when it comes to learning . This method uses the Farnsworth system. This system increases the spacing between letters (slows down the speed between characters).

In the 1930s, Ludwig Koch, a German psychologist, developed an easy method for it to be quick and effective to train the many commercial operators needed at that time. He investigated how they had taught the best active operators. Reaching the conclusion, that three quarters learned through sound and the rest in writing (cards with the printed code).

Listening to the code at a high speed from the beginning lets us recognize every character as a single sound, like a little melody we record in our brain.

After many practical tests, Koch realized **that operators who had learnt only through sound, could decode the code at high speed. On the other hand, the ones who used printed card, decoded at low speed**. This is because they had to do the "translation" of the sound to the image and then recognize it. These middle steps slow you down.

At first it may seem that the order of the characters introduced in each lesson of the Koch method is completely random. However, it is designed to help you gradually recognize the "Morse rhythm" and differentiate similar sounding patterns and the more difficult or "annoying" letters.

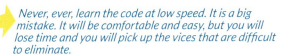

Never, ever, learn the code at low speed. It is a big mistake. It will be comfortable and easy, but you will lose time and you will pick up the vices that are difficult to eliminate.

- **COMPUTER PROGRAMS**

G4FON TRAINER - 8/10

www.g4fon.net

- Koch method.
- Good trainer that simulates interferences, noise...

MORSE MACHINE - 6/10

www.g4ilo.com

- Koch method.
- Not a great user environment, but still works.

JUST LEARN MORSE CODE - 8/10

www.justlearnmorsecode.com

- Koch method.
- Easy to use.

- **MOBILE APPS...**

IZ2UUF CW - Score: 8/10

www.iz2uuf.net/wp/index.php/2012/07/02/koch

- Perfect to use it on your android mobile phone.

MORSE MACHINE APP - 7/10

https://play.google.com/store/apps/details?id=com.iu4apc.morsemachine

- Addictive on your android mobile phone.

HAM MORSE - AA9PW - 8/10

https://aa9pw.com/ham-morse

- iOS, very complete application.
- Koch method.

FARNSWORTH* SYSTEM, VISUAL EXAMPLE:
Real speed 20wpm, spacing between characters 20wpm (20/20 wpm).

▬ ▬ ▪ ▬ — ▪ ▪ ▪ — ▬ ▬ ▬

*Real speed 20wpm, spacing between characters 10wpm (20/10 wpm).
20/10wpm has double spacing between characters.

▬ ▬ ▪ ▬ — — ▪ ▪ ▪ — — ▬ ▬ ▬

Note.- Further information about spacing and speed on pages 24-25.

1 Set up the program you are using to a real speed of 20wpm and a 10wpm spacing (better if the value is over 10).

2 Set the length of every lesson up to at least three minutes and practice 20 or 30 minutes per day, EVERYDAY, until you finish them.

INDIVIDUAL TRAINING METHOD FOR LEARNING MORSE CODE

Follow the steps of the image as well as the ADVICE and TIPS from the next chapter.

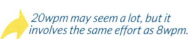
20wpm may seem a lot, but it involves the same effort as 8wpm.

With this method of individual training, you will learn at your own pace and take the time needed for each letter. Some may cost you more than others, but you should not give up, remember that the only trick is to have:

DETERMINATION and PATIENCE

When you start a new lesson typically your success rate will drop. Do not worry, keep trying until you reach at least 90%, that is the time to move on to the next one. If you are constant, I assure you that you will be able to learn the code in a few weeks.

You are learning at 20 wpm!
If you have followed the instructions, remember that you have begun to receive correctly at that incredible speed. Congratulations !!

3 In the first lesson you will be given two letters in groups of five.

Listen and write them down in a notebook or computer.

Koch method, first groups – some examples:

KMMKM KMMKM KMMKM
KMKMM KMMMK MKKMM

Advanced groups – some examples:

KJH56   H?KTT   /HZYC
JL=O4   18K9R   OQ.T?

4 Once the lesson is finished, check for mistakes. If you are over 90% correct, move forward to the next lesson (a new character will be added).

After teaching various morse code courses as a monitor, I have been able to verify that this system is the one that achieves the best results, in the shortest time.

If you also want to achieve it, you just have to limit yourself to following the suggestions that follow.

TIPS AND TRICKS
to study and learn the code

> *Please read the following pages carefully, they will help you a lot.*

If you talk to any radio telegraphist about what the trick is, what are the most basic and important things when learning the code and making contacts, everyone will tell you:

LISTEN, LISTEN AND LISTEN AGAIN

You need to understand and learn the sound pattern of every letter, number or symbol. Without thinking and of course never counting the number of dots and dashes heard or using mnemonic rules of words that resemble the sound of letters.

Remember that with PRACTICE, we will be able to strengthen in our subconscious the reflexive act of assimilating a sound into a character. That is our objective.

The first step is going to LCWO.net and signing up or downloading any of the apps from the previous chapter. You are free to practice using the method you prefer, but always use a good quality code source and, now that you are learning, with a tone that is pleasant to you.

When you are on the radio, the tone could be surrounded by noise and interference and even fade, etc. While that you are learning you should practice with a tone that is comfortable for you (the usual thing is to configure your application to sound between 500 and 1,000 Hz).

Now let's see some tips and tricks that may be helpful.

But remember **the key** is patience.

BE CONSISTENT AND PATIENT

about the lessons

- First of all remember that we are practicing something that is our hobby, **enjoy the experience, do not get overwhelmed**.

- **If you start at a high character speed, it will stop you counting the dots and dashes.** You will recognize every character as a sound unit. It is about "recording" each sound as a reflection in your mind, without counting, without thinking.

- As we saw in the last chapter, **try setting the program up to 20wpm with a 10wpm speed, so that the spacing between each letter is bigger and it will be easier to distinguish them**. Once your control is good, start reducing that difference and practice with the same real speed (20/20, 24/24...).

- Listen to the lessons, groups of characters, for no more than 30 minutes per day. If you use the KOCH method, you will start with letters M and K. The rest will be added afterwards.

- If you have enough time, try a second session. Don't worry if you get stuck. Try to space one lesson from another. You may make rapid progress at the beginning and then you will get stuck. Don't worry, it's normal.

- **Don't move forward if you find problems with a lesson.** Keep trying, repeat it as many times as necessary. If your score is 90% or more, it is time to continue.

- **FEEL THE UNIQUE SOUND - ITS MUSIC -** , use your intuition to decode each character and write it down. Don't worry too much and **DON'T THINK** whether it is right or not. Focus and decode without thinking.

- Once written, get ready for the next one. **If you couldn't recognize it, leave a mark and wait for the next one. This is essential. Never get stuck on a missed letter, try to focus on the following ones.** The more speed, the more characters you will miss. I know this may be difficult at first, but you must get used to letting go of the letter and forgetting it. Letting a missed letter pass is essential when making radio contacts.

- Never see the lessons before you listen to them.

- **If you are stuck for many days on a lesson, take some rest.** After a break, you will see things clearer. It may seem tough, but it is worth it.

some advice

> Never have a printed sheet with the dots and dashes of each letter in front of you. Forget that idea and remember that you are learning SOUNDS.

- What might confuse you is not the speed of the letters but how often they arrive. With practice you will overcome any difficulty.

The most used symbols are:
? / =

- Practicing for too long each day may not help you make progress. There comes a time when our mind "gets tired".

- At first you may find it difficult to write what you receive in a notebook or on a keyboard due to the speed but believe me, you will overcome this in time. There is no need to write with perfect penmanship, adapt the letters, mix uppercase and lowercase letters and deform your writing if necessary.

- Each character is unique and different, do not mix or compare them. Do not visualize "mirror" letters or numbers and avoid making comparisons that slow down learning and translating. Don't be fooled, these tricks are not useful.

- Everyone has their difficult characters which at first appear to sound similar. Programs and applications allow you to create character groups where you can select your challenging characters and practice them in isolation, repeating but never slowing down.

- Practice anywhere. Walking down the street, translating car license plates, store names to DITs and DAHs...

- Just starting out, the day will come when you stop writing everything down and just capture the key information for the QSO. But for now fill sheets of paper or write down the groups of letters on the computer and do it again and again.

- Remember that what you are hearing you will have to "write" when it comes to sending. That is why it is very important that you establish very clear sound patterns in your mind.

- Later on you can generate MP3 audio files with text from books, even in a language you do not know. If you cannot predict the words the practice will be more effective.

- Whenever possible, listen to the CW on your radio wearing headphones. You will realize that you will hear many more details than with the speaker.

- Don't look for excuses to stop practicing, you have to be patient. Even if it's harder than you thought at the beginning, you have to be persistent and not give up.

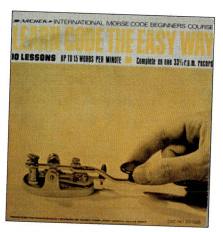
Vinyl course 1962?

- Some programs or applications allow you to try listening to calls, Q codes, abbreviations, contest formats including adding background noises, etc. When you know the code they will help you practice more realistically.

- Even when you know a character, you may unintentionally try to COUNT the dots and dashes. If you catch yourself doing this, you must correct yourself immediately. Remember, do not visualize dots or dashes or count them. Doing so will slow down your learning.

- It is interesting to practice numbers separately. Configure your application for it and make groups of just digits. It will be very useful additional help.

- With time and practice you will find you can listen to the code without paying too much attention, clearly understanding what has been received.

- After practicing, turn on the radio and listen to CW for a few minutes around the indicated frequencies. Suddenly everything is different, the letters run together and it seems that everyone is in a hurry. Relax, the trick is to LISTEN. At first you won't understand anything, but you will hear the "music" that will be repeated again and again. For now, don't worry too much, these are small steps.

- Almost all bands have a frequency for novice radio amateurs to meet and work at low speed or QRS and those experienced wanting to help practice with them (see table on the right).

- Don't give up on the first attempt. If you don't do well, keep trying. Only through PERSERVERENCE will you learn. Finding excuses is very easy: work, family ... It took me so long to learn that I am ashamed to say (I quit more than once, big mistake).

CW QRS Frequencies
10m 28.055 MHz
12m 24.905 MHz
15m 21.055 MHz
17m 18.085 MHz
20m 14.055 MHz
30m 10.125 MHz
40m 7.035 MHz
80m 3.555 MHz

DON'T MAKE EXCUSES, MAKE IMPROVEMENTS

- If during the lessons you have learned with a real speed of 20 wpm, you are perfectly qualified to start operating on the radio at 14 wpm or even more.
- At the moment of practicing sending (p. 56) you can use the free android App "MORSE CODE READER". This App decodes your transmission and tells you if you are making the dots, dashes and spaces correctly.
- **It is normal to get nervous** during the first QSO. Don´t worry, all radio hams have had that problem at least once.
- If you can make your first QSO with your radio ham friends you will fell less nervous and they can help you correcting your mistakes.
- **Record your first QSOs and listen to them** after you have overcome your nervousness. Find your own mistakes and correct them.
- **Create MP3 files** with your information (callsign, name, city...) and set them as a ringtone. This way you practice.
- Use your morse code practices as **an escape** from everyday life. Take some minutes to practice and break away. If you are not enyoing your study, leave it for another day.
- **Motivation is the main tool that you will need to learn the code.** Take advantage of all the materials you are provided with. Anytime, anywhere is good.

| It is important to practice beyond your posabilities (faster than you find comfortable). I know it's hard, but it's the best way to learn. | If you start working at a real 20 wpm speed but you see no improvement, slow to 18 wpm, 16 wpm... (but first you must try seriously). |

- Don´t get frustrated, the important thing is to learn and we all have our own pace. -

MFJ-564

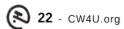

22 - CW4U.org

CW or MORSE CODE
Are they the same?

IT'S ONE THING THE CODE, IT'S ANOTHER THE RADIO TRANSMISSION

If you are a radio ham, **you already know that CW, morse code and telegraphy are quite different** even though we interchange the terms. I shall briefly explain the difference to those who don't know.

The coding of letters and signs in dots, spaces and dashes was called **MORSE CODE** (p.10). At the beginning, the code was sent as pulsed electric current through a wire between two or more points (telegraph). **CW is the way to transmit those pulses by radio** (radiotelegraphy).

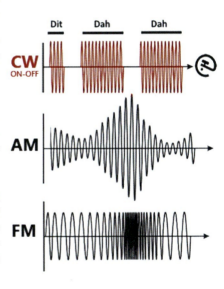

How do we get that broadcast? Through **radio frequency waves** that can be transformed, by modulating a **carrier signal**.

Depending on how we modify that carrier wave, we can send a frequency modulated signal (FM), an amplitude modulated signal (AM) or an unmodulated **CONTINUOUS WAVE (CW)** where we send or interrupt that carrier.

The cutoff and return of the transmission and therefore its duration is done by hand or electronically with a **switch** known as a key.

SIGNAL MODULATION

The popularity of CW among radio hams is due to its signal/noise ratio, reduced bandwidth and therefore a great transmission performance, making it easier to make long distance contacts even in unfavourable conditions.

The first system of communication for sale was Guillermo Marconi's, although the first patent was by Nikola Tesla.

CW4U.org - 23

RECOGNIZE
the sounds, space and timing

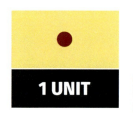

1 UNIT
DOT

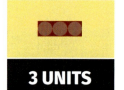

3 UNITS
DASH

"The pace is the code's soul. There is no morse without it."

Morse code "translates" each character into a different combination of long and short sounds and their length and separation are differentiated by **measuring their timing**.

That timing, that unit of measure, results from the LENGTH of the sound of the DOT, defining this length in the timing of the dashes and spaces.

With practice you will be able to mentally measure the times, creating in your transmission a music of measured and precise rhythm. An irregular rhythm hinders the process. Spacing is important, as well as the formation of the characters, especially if you send very quickly.

In the next image, a line of dots and dashes, you have a representation of the timing between words and letters in a text.

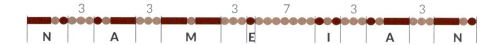

1 UNIT
Separation in a letter

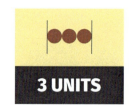

3 UNITS
Letter spacing

7 UNITS
Word spacing

WORDS PER MINUTE
accuracy transcends speed

YOU SHALL NOT BE OBSESSED WITH THE SPEED.
The main thing is that what you hear or transmit is perfectly recognizable.

The Morse code send speed is measured according to **words per minute (wpm)**. The abbreviation **cpm**, characters per minute, is also used.

For this measure the word PARIS is taken as a reference and the number of times it is sent in one minute. This way, if you can broadcast this word twelve times in 60 seconds, your speed will be 12 wpm.

You should already know how important practice is, as well as listening before transmitting. Once you recognize sound patterns, you can send at the speed you feel comfortable. The speed can be increased as you gain experience.

About the speed.-

- Never send beyond the speed you can receive when making contacts on the radio, maintain distances between dots and dashes (correct distance).

- When we start we all go slow, it is not a problem, there are many like you and expert radio partners who will lower their speed to match yours.

- The main goal is the code to be as correct as possible.

- If you are learning, when having a QSO ask the other station to QRS (see p. 72).

- Never learn Morse code at low speed **(it's a trap)**.

PARIS has a length of 50 units of time (t).

P di da da di = 1 1 3 1 3 1 1 (+3) = 14 t
A di da = 1 1 3 (+3) = 8 t
R di da di = 1 1 3 1 1 (+3) = 10t
I di di = 1 1 1 (+3) = 6 t
S di di di = 1 1 1 1 1 [+7] = 12t

TOTAL: 50 units of time
(+t) = time between characters
[+t] = word spacing.

In May 2003 Andrei Bindasov EU7K1 (Belorussia) successfully sent 216 characters in one minute.

Once you have learnt to use Morse code, we have to know how to make contacts on the radio. In order to do that we need to know some codes, abbreviations and compositions for the QSO. We will see that in the following pages.

ETIQUETTE
rules, codes and patterns

Telegraphy is similar to driving. If you just follow the content of this book, you will only learn the theoretical part. We also need to have some practical lessons. Can you imagine going out on a highway without knowing how to drive and ignoring traffic rules?

It is important that, while studying the code, we also listen to CW on the radio. At first everything will sound very strange and give the impression that everyone transmits nonsense. But little by little, as we move forward with our classes, you will discover that some rules are usually followed when making a QSO. The data is given in an order, using codes, prosigns and abbreviations to shorten the sentences.

Since the beginning of the telegraph, abbreviating was meant to be the easiest and most profitable way of working. It was frustrating and tedious to send lots of characters, that is why abbreviations and codes were created. Today, radio hams still use those "shortcuts".

What's the point of having a complex and erratic conversation through dots and dashes instead of using a simplified model? **Better if a previously established order is followed, so we will reduce errors, especially when receiving.**

That doesn't mean that we cannot have a fluent conversation in CW. There are several types of QSO; on the one hand we have the brief contact (in a contest) and on the other hand a usual contact (DX, friendly chatting).

As I have already said, the fundamental thing more than the speed, is that the code that we listen to or transmit is as perfect as possible and if we also operate following the code of conduct and the guidelines that we are going to see, nobody can tell us that we are doing it wrong.

If you are an experienced radio ham with experience in phone you will need to change some concepts, even what you think the meaning is of a code whose initial use was in CW.

There are plenty of abbreviations and forms of Q code. You will find some of them at the end of the book. For the moment, **you only have to learn the most commonly used to make contacts in CW**. They are detailed on the next few pages.

When commercial telegraphists came out, they used a book on abbreviations called "Phillips Code", that shortened the text up to 40%. It had over 6000 abbreviations.

Q codes
Many things with only three letters

This code always starts with the letter Q. In the first decade of the 20th century, the code was used in ships and coast radio stations. **They not only shortened the transmission, but also made communications easier** and clearer between naval operators from different nationalities. Today they are used by radio hams and aeronautical communications.

In telegraphy the Q code is indispensable. You must learn it.

NEVER write **ES** or **IS** between the code and the associated information.

The Q code can go alone or followed by a question mark (QTH?), in which case its meaning refers to a question. In the table below, you have some of the most common and at the end of book another longer list.

	INFORMATION	QUESTION (?)
QRA	This is ...	What is the call sign of your station?
QRG	Your exact frequency is ...	What is your frequency?
QRK	The intelligibility of your signals	Will you tell me my exact freq?
QRL	I am busy (or I am busy with ...)	Are you busy?
QRM	I am being interfered with ...	Are you being interfered with... ?
QRN	I am troubled by static	Are you troubled by static?
QRO	Increase transmitter power	Shall I increase transmitter power?
QRP	Decrease transmitter power	Shall I decrease transmitter power?
QRQ	Send faster (... wpm)	Shall I send faster?
QRS	Send more slowly (... wpm)	Shall I send more slowly?
QRT	Stop sending	Shall I stop sending?
QRU	I have nothing for you.	Have you anything for me?
QRV	I am ready	Are you ready?
QRX	I will call you again at ... hours	When will you call me again?
QRZ	You are being called by ... on ...	Who is calling me?
QSB	Your signals are fading	Are my signals fading?
QSL	I am acknowledging receipt.	Can you acknowledge receipt?
QSO	I can communicate with ...	Can you communicate with ...
QSY	Change to trx on another freq...	Shall I change to trx on another freq.?
QTH	My position is ...	What is your position?

Abbreviations
The economy of language

The most common words, verbs, etc, have always been abbreviated to make transmission easier, clearer and shorter and to avoid interferences problems, fading etc., and also not to lengthen the writing time of the code too much. It's not so rare, today we do the same with *SMS* or *Whatsapp*.

Normally, we use abbreviations of English words and the majority of them were created by former telegraphists. Some abbreviations include removing vowels, use of phonetic symbols, etc.

Here you have some of the most common abbreviations in CW. See Annexe for more.

AGN	Again		**NR**	Number - Near
ANT	Antenna		**NIL**	Nothing. Not in log
BK	Break, Break in		**GG**	Going
BURO	Bureau		**OM**	Old Man
B4	Before		**OP**	Operator
C	Correct, yes, copy		**PWR**	Power
CL	Call - Close		**PSE**	Please
CONDX	Conditions		**R**	Roger, **OK, C**
CPI	Copy		**RPT**	Repeat
CU	See you		**RPRT**	Report
CUAGN	See You Again		**RIG**	Station equipment
DE	This Is, from		**RST**	Signal report format
DR	Dear (DR OM)		**SRI**	Sorry
ES	And (GL ES DX)		**TEMP**	Temperature (F or C)
FB	Fine business		**TNX**	Thanks (**TKS**)
FER	For... (**FR, 4**)		**TU**	Thank you (*in the end*)
GB	Good bye, God Bless		**U**	You
HI	The telegraph laugh; High		**UR**	Your
HR	Here; Hear		**VY**	Very
HW?	How, How Copy?		**WX**	Weather
N	No, Negative, Incorrect		**73**	Best regards

Prosigns
Procedural signs

Except for K, **they are created by two letters transmitted together without spacing,** creating a unique sound.

They are very useful within the conversation **and a reminder that we are at a certain point of the QSO that needs a specific action or procedure**, like changes in the transmission for instance.

Prosigns are normally written between the sym<u>bo</u>ls **< >** or with a line on top (**<AR>** , $\overline{AR}$).

It is very important to use them correctly.

These are the most used prosigns *(procedural signs)* during CW contacts.

< K >	Open exchange

- Invitation to transmit after terminating the call signal.
- CQ CQ CQ DE EA7HYD EA7HYD <K>

< KN >	Exchange for you

- Invitation for named station to transmit. Normally with which the QSO is being performed.
- ... HW DR MIKEL? EA2CW DE EA7HYD <KN>

< BK >	Fast exchange

- Abbreviation for "back-to you". Break in conversation.
- <BK> SRI QRM QRM HR PSE RPT <BK>

< AR >	End of message
	· <AR> is sent when the message is finished, and you pass your turn. · It is sent at the end of the transmission and before the callsings. · ... HW? <AR> EA2CW DE EA7HYD <KN>

< SK >	End of contact, also <VA>
	· <SK> indicates that this will be the last transmission of QSO. The station that started the QSO is normally the one to end it. · There is no transmission after <SK>. · ... 73 ES GUD DX <BT> EA2CW DE EA7HYD <SK>

< BT >	Space or stop. Also represented with the equal sign =
	· It is used to separate the different parts of the QSO. If we send one after the other we indicate "hold on a moment, I'm thinking." · ... TNX FER CALL <BT> UR RST 599 5NN <BT> NAME IS DAVID DAVID ...

< AS >	Wait, please. Make a pause.
	· We ask someone to wait because we are busy. · We make a stop to adjust the station, write something down... · .... NAME DAVID DAVID <AS> · · · · · · · · QTH NR CADIZ CADIZ =....

< SN >	Received, also <VE>, R, OK
	· It is used to express agreement or to tell that you have successfully received the message. Understood. **Ve**rified. · ... <SN> ALL FB QSO ...

In some publications they recognize as prosigno:

< HH >	Error
	· The same meaning has three points somewhat more separate from normal *(E E E)*, after them we will repeat the entire wrong word. · ... QTH NR CAB <HH> CADIZ ...

Prosigns have been used since the 1860s until today. At first, there were twelve, but some of them are no longer used. Prosigns generate **very characteristic sounds** that you must learn and differentiate in the QSO.

Signal report
RST a method of reporting radio signal quality

We all want to know the quality of our signal because it is a very important piece of information in a QSO. In order to do that we use a report following the RST format.

In order to do that we use a three digit number, with one digit each for conveying an assessment of the signal's readability, strength, and tone, which is measured with the following scales:

READABILITY	**S**TRENGTH	**T**ONE
1.- Unreadable	1.- Signals barely perceptible	1.- Extrem. rough hissing note
2.- Barely readable	2.- Very weak signals	2.- Rough AC note, no musicality
3.- Readable with difficulty	3.- Weak signals	3.- Rough, low AC, slightly musical
4.- Readable	4.- Fair signals	4.- Rough AC note, dry musical
5.- Perfectly readable	5.- Fairly good signals	5.- Musically modulated note
	6.- Good signals	6.- Modulated, slight trace of whistle
	7.- Moderately strong signals	7.- Near DC note, smooth ripple
	8.- Strong signals	8.- Good DC note, just a trace of ripple
	9.- Extremely strong signals	9.- Purest DC note

Actually, **the tone (T)** was much more important at the beginning of amateur radio, since the power supply and the equipment itself did not have the current quality, producing ups and downs, sparks or displacements.

The **R** and **T** is based on a subjective opinion upon how we receive a signal with the information above.

The signal strength (S) can be measured by the S-meter. Its sensitivity must be calibrated under certain criterion. In commercial radio stations they come with a default setting. On the internet you can find decibel tables for each S. Example: S9 = 56 dB, 50 uV, - 73 dB.

In contests you may pass a report of 599 or 5NN. Contestants make sure that information is correct. But, in a normal QSO, you must pass the report as you actually hear them. This is the way the speaker will know about the quality of the signal. I always work with transceivers and self-built aerials, and I wait for the real RST and look forward to seeing how my signal was received.

> Numbers can also be abbreviated. The most common numbers are:
>
> *5 = E , 9 = N , 0 = T*

Other usual reports can be: 559 or 579.

Information order
QSOs do have a structure

When making a contact on the radio through Morse code you **must follow an established order. That way we make sure that the necessary information is correctly sent and received** to record the QSO in our log book or to submit a log for a contest.

We must differentiate the time used (short) and information sent (succinct) in a QSO with a contest or a DX station compared to a contact that can be made in a relaxed way from our shack.

Many things can be added in every exchange, for instance, asking if the signal was clear and/or correct (HW?). In a normal contact of CW, information is normally arranged like this:

ORDER OF INFORMATION - BASIC QSO

1 — CALL SIGNS
After the call or answer, callsings are confirmed

**2 — GREETING
THANKS FOR ANSWERING
REPORT, NAME, QTH**
-are the minimal information-

**3 — OPTIONAL.-
EQUIPMENT, POWER, ANTENNA
WEATHER, TEMPERATURE ...**
We can extend and send more information, establish conversation...

4 — we confirm or ask for a QSL
FAREWELL
WE EXPRESS OUR THANKS FOR THE QSO AND **FINISH** the contact

> In the following pages we are going to see how to make a QSO on the radio, some of the most frequent doubts and pitfalls and how to face them. We will also see how to use a key and how to make contacts in contests, pile-ups, etc.

MAKING A QSO
real contacts in morse code

This is the most interesting part: how to make contacts on the radio. In order to explain it step by step, I have made a basic QSO template using a 'WhatsApp conversation' as a reference. Each exchange is labelled with a number, that way we can locate the part of the QSO we are referring to and on which we want to comment.

An example of a contact found in books would be written like this:

QRL? CQ CQ CQ DE XX1AAA XX1AAA XX1AAA K

XX1AAA DE ZZ2HH ZZ2HH AR

ZZ2HH DE XX1AAA TNX FER CALL BT UR RST 599 599 HR QTH CITY CITY BT NAME IAN IAN SO HW CPI? ZZ2HH DE XX1AAA KN

XX1AAA DE ZZ2HH TNX FER RPRT BT SLD CPI UR RST 589 589 BT NAME JOAN JOAN BT QTH TOWN TOWN BT HW CPI? XX1AAA DE ZZ2HH KN

ZZ2HH DE XX1AAA TNX FER FB QSO JOAN BT HP CU AGN BT VY 73 TU ES URS SK ZZ2HH DE XX1AAA

This would be one of the many so-called "template QSOs" and, apart from the prosigns not being enclosed between the < > symbols (I've underline them), it is well written and correct. Today I can read and understand it but, when I began to study how to make contacts in Morse, I found it really difficult. I knew the prosigns, Q code and abbreviations, but...why did each QSO seem to be the same as the previous one but with the text changed? What strange meaning was there among so many letters? What degree of madness is necessary to repeat them like this, again and again?

Real contacts in Morse code
example of a possible QSO between two stations (p. 1/2)

EA7HYD

OPERATOR.- David
LOCATION.- Chiclana de la Frontera, 18 Km from Cádiz, SP.
EQUIPMENT.- Self-built SDR.
POWER.- 10 watts
AERIAL.- dipole
QSL.- direct or eQSL.
+info: www.cw4u.org

This is a practical example of a QSO.
In the following pages, we are going to explain and comment on every point step by step.
www.CW4U.ORG

❶
QRL?   QRL? DE EA7HYD

CQ CQ CQ DE EA7HYD EA7HYD PSE <K>

CQ CQ CQ DE EA7HYD EA7HYD <K>

❷ EA7?

❸ DE EA7HYD EA7HYD <K>

❹ EA7HYD DE EA2CW EA2CW PSE <KN>

❺
EA2CW DE EA7HYD GD DR OM
ES TNX FER CALL =
UR RST 599 5NN =
NAME IS DAVID DAVID =
QTH NR CADIZ CADIZ =
HW? EA2CW DE EA7HYD <KN>

---

❶ We ask whether the frequency is free or not.
· We make an open exchange <K> waiting for an answer.
· We wait a few minutes for an answer (~~~~).
· We make a general call a few times.

❷ Someone has heard our call and wants to make contact, but he or she hasn't heard it completely.

❸ We repeat our callsign.

❹ Our icontact now has our QRA and has slowed down their sending speed to match ours.
· Close with <KN>, because the exchange is now with a 'named' station.

❺ First, we send his callsign then ours and we say, "good day, dear friend and thanks for the call".
· We send the RST reporting how well we hear him, our name and location, near Cádiz.
· We ask him how he is receiving us, and we hand back to him with <KN>.
· We separate the information with the prosign <BT> also written =.

In the following pages we analyse this QSO in-depth

(p. 2/2)

EA2CW

OPERATOR.- Mikel
LOCATION.- Bilbao
EQUIPMENT.- Youkits HB-1B
POWER.- 10 watts
AERIAL.- dipole 7-28 MHz and tuner.
QSL.- Bureau.

DOWNLOAD for free the mp3 file and other QSOs from the website: www.cw4u.org

❻
EA7HYD DE EA2CW GM DR OM DAVID =
UR RST 579 57N =
OP MIKEL MIKEL =
QTH BILBAO BILBAO =
HW? EA7HYD DE EA2CW <KN>

❼
R R EA2CW DE EA7HYD =
TNX FER FB QSO DR OM MIKEL =
73 ES GUD DX = HPE CUAGN <AR>
EA2CW DE EA7HYD <SK> TU

❽
EA7HYD DE EA2CW OK DAVID =
TKS FER NICE QSO ES QSL VIA BURO =
73 EA7HYD DE EA2CW <SK> TU E E

❾
E E
CQ CQ CQ DE EA7HYD EA7HYD <K>

❻ He answers with our callsing and name, says hello, sends our report and then tells us his name and location.
· No more information is provided.
· He asks how we copy him then passes the exchange back to me with <KN>

❼ We let him know that we received everything correctly (R R).
· We start with the callsigns again and thank him.
· We send our farewells, wish him good DX, hope to see him again and indicate this is the end of the message with <AR>.
· We give callsigns and close the QSO with <SK> and say thank you.

❽ He says goodbye and thanks for the QSO. He also tells us about the way of sending a QSL.
· Farewells, callsigns, contact close, thank you and two E.

❾ An informal goodbye would have two dots.
· We start calling again, without asking whether the frequency is free or not because we are already using it.

CW4U.org - 37

Analysing a QSO
What else could have been said or done?

unknown key

> The first thing you need to know is that the previous example of a QSO was more practical than theoretical. I apologise to the purist, but I am going explain and follow the rules of the real and regular contacts.

I guess there are as many ways to make a QSO as there are operators. We are going to use the QSO from pages 36-37 as an example to see how it was done and what kind of information operators could have exchanged, the variations and problems that could have appeared.

> **Each exchange of the QSO and lines are numbered.**
> For instance, ❶ 2 = first exchange, second line
> (this will make it easier to comment on it in the book or when discussing it with radio colleagues)

The example has been divided into the following sections (exchanges) to explain the possible parts of a QSO in CW. This doesn't mean that all CW must be like this, but the most common exchanges are represented.

EXCHANGE	ACTION
❶	**S**TARTING QSO
❷-❸	They haven't heard our callsign
❹	**W**E ANSWER A CQ OR RECEIVE AN ANSWER
❺	**E**STABLISHED CONTACT, WE ADD SOME INFORMATION
❻	**W**E CONTINUE AND ANSWER
❻+	... extending the QSO, additional information
❼-❽-❾	**F**AREWELL AND CLOSURE

❶ Starting a QSO

1 - QRL?   QRL? DE EA7HYD
2 - CQ CQ CQ DE EA7HYD EA7HYD PSE <K>
3 - CQ CQ CQ DE EA7HYD EA7HYD <AR><K>

Let's suppose that **we want to make a call** on any band. In order to do that we go to the segment that is often used for telegraphy in that band (1* - *see notes at the end of every section*) and after checking the frequency is free and adjusting our equipment, we follow these steps:

First of all, we listen for a few seconds (or minutes) to see whether the frequency where we have decided to call is free (2*).

❶1.-   **QRL?**   we ask once if the frequency is busy. We don't even give our callsign to minimise disturbance to other stations. We resume after a while.

QRL? DE EA7HYD   we ask again. If the frequency was busy they could answer: QRL would be perfect, but to save time we use: E, EE, C, R, Y, QSY. We won't answer. We can also receive <AR> or even QRX to make us wait.

If the frequency is busy, we will search for another, but, if we don't get an answer, knowing the frequency is free, we make the call:

❶2.-   **CQ CQ CQ DE EA7HYD EA7HYD PSE <K>**   we send CQ three times and our callsign twice, no more than that (3*). We end with "please" (PSE) meaning "*please answer*" and we leave the exchange open <K>, inviting responses.
We wait for a minute and call again as many times as we want.

❶3.-   **CQ CQ CQ DE EA7HYD EA7HYD <AR> <K>**   we indicate the end of the message <AR> and leave the exchange open to any station <K>. We could even end simply with <AR>.

IN EXCHANGES OUR CALLSIGN GOES AFTER THE OTHER STATIONS

VARIATIONS ❶

After each exchange of the example QSO, some of the variations that could have been used are indicated. Of course there could be many, many more.

❶ .-   **VVV VVV   DE EA7HYD EA7HYD**   was used in the early days of radiotelegraphy to carry out transmission tests and some operators continue to do it ... although sometimes they are not testing anything and are actually calling.

> 1* .- If we are inexperienced it is best to place ourselves close to the QRS calling frequency or near the top of the CW segment of the band in which we are working. The lower part of the CW segment is where you will find the more experienced operators. If we are operating on the weekend of a competition, it is possible that the whole band is busy so it is best to listen (for now).
>
> 2* .- Even if we are not hearing anything we should ask if the frequency is clear, because it is possible that a QSO is being made and we can't hear the operator who's sending, but he can hear us or his QSO partner can. In that case, we risk disturbing their QSO.
>
> 3* .- Sometimes we listen to an infinite number of CQ calls, but then, we just hear the callsign once (the most important thing).

My teacher (EA2CW) repeated this phrase to me:

> **"A QSO is a dance between two people who know the music, the rhythm and the steps to take"**

... try not to step on anyone's toes.

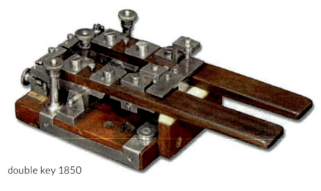

double key 1850

❷ ❸ They have not heard our callsign

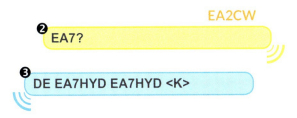

The callsign of the two stations is the main part of the contact, without this information the QSO would not make sense.

❷.- EA7?   Someone has heard us but couldn't copy our callsign correctly. It may be missing the last letter or two but, rather than asking the long and complicated question, just send EA7?

❸.- DE EA7HYD EA7HYD <K>   we repeat our callsign once or twice and send an open exchange to any station <K>. As we can see, in this exchange we don't send CQ and simply send the DE directly.
We will answer and repeat the callsign as many times as necessary until the two QRAs are clear.

VARIATIONS ❷ ❸

❷1.- EA7 CL?   they send the beginning of our callsign and ask us who is calling (CL?).

❷2.- EA7? DE EA2CW <K>   much better as the station that answers gives their callsign (DE EA2CW) and so EA7HYD knows who their QSO partner is. The exchange is left open <K> as it is still not known if the communication has been established.

❷3.- QRZ? DE EA2CW <K>   we indicate that we have not understood the callsign (QRZ?) Followed by our QRA and an open exchange <K>.

> **Note.-** You must keep in mind that the transmission in point 2 is illegal as the station has not identified itself but, in practice, if you hear a question mark (?) or part of your callsign followed by a question mark after making a CQ call, it is obvious that someone is asking you to repeat your call. This is common practice when making contacts in CW.

❹ We answer a CQ or receive a reply

❹ EA7HYD DE EA2CW PSE <KN>   EA2CW

The way to start a contact is to call CQ, **answer a CQ or listen to a QSO in progress and wait for it to end**. We will never call while the contact is in progress.

- If what we hear is a CQ we will copy the QRA and answer the call with it.

- If we are waiting for the end of a QSO that is in progress, we can already be collecting data on the callsign, name, location, etc., so we will have part of the station´s information before starting.

In the example provided we would call the station heard with its QRA and end with <KN>, as in point 4 of the QSO, where EA2CW already has our callsign and answers our call:

❹1.-   **EA7HYD DE EA2CW PSE <KN>**   EA2CW ask us to please (PSE) respond and passes control back with a closed exchange (<KN>). This indicates that the change is only for EA7HYD, since it is with whom you try to make contact.

VARIATIONS ❹

❹1A.-   **EA7HYD DE EA2CW EA2CW <AR>**
we end the message waiting for a reply (<AR>).
As we don't know if EA7HYD will answer,
we don't end the exchange with <KN>.

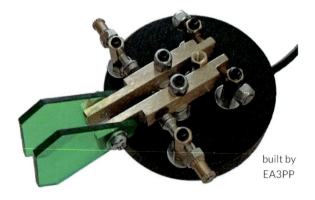

built by EA3PP

❺ Contact established, we provide our data

1.- ❺ EA2CW DE EA7HYD GD DR OM
2.- ES TNX FER CALL =
3.- UR RST 599 5NN =
4.- NAME IS DAVID DAVID =
5.- QTH NR CADIZ CADIZ =
6.- HW? EA2CW DE EA7HYD <KN>

They heard us and we have the callsign of the station with whom **we can start having a QSO**, sending its signal report and our personal information.

Let's look line by line at the text of the transmission in section 5:

❺1.- **EA2CW DE EA7HYD GD DR OM**   from this moment and since we know the QRA of both stations, any exchange must begin and end with the callsign to whom the message is addressed and the callsign of the sender, always separated by DE. We continue to wish a good day (GD) colleague, or old friend (OM).

❺2.- **ES TNX FER CALL =**   and (ES) thanks (TNX) for calling. Then we send <BT> (=) thereby making a separation or a break between sections of the text (1*).

❺3.- **UR RST 599 5NN =**   your (UR) signal report (RST) is XXX. We report the signal as we saw on page 32.
When we repeat the report a second time we can abbreviate the numbers: 599 = 5NN (2*).
Again, as it will become customary, we separate this information from the following lines with the prosign <BT> which represents a = or 'break'.

❺4.- **NAME IS DAVID DAVID =**   the operator name is normally sent at least twice (3*) and again we leave a break (=).

❺5.- **QTH NR CADIZ CADIZ =**   we send our location (QTH), also repeating the name of the town twice.

As we can see in the template information, the name of the operator´s town is Chiclana de la Frontera. The name is a bit long, so it's better to say that he lives near (NR) Cadiz, a much shorter name that's quicker to send. Again we send a break between information (=).

❺6.-   **HW? EA2CW DE EA7HYD <KN>**   we ask if he has received us OK (HW?), and we give the callsigns again and pass the exchange back to the other station (<KN>).

VARIATIONS ❺

❺4.-   **OP DAVID DAVID**   or alternatively   **MY NAME IS DAVID DAVID**   are other ways of giving our name.

❺6.-   **<AR> EA2CW DE EA7HY <KN>**   the prosign <AR> can also be placed at the end of the message but always before the callsigns.

If, during the QSO, we do not copy the name or some other information we can request that they repeat it for us. For example:

<BK> PSE UR NAME ? <BK>   please (PSE), your (UR) name? (NAME ?) or

<BK> PSE RPT NAME <BK>   please (PSE) repeat (RPT) name (NAME).

The use of the quick exchange **<BK>** is found on page 54 and in the annex.

1*.- Some operators do not use the prosign <BT> or = to take a break or separate information, instead they use the full stop or period (AAA). Using it is the most practical and useful method and the listener will appreciate it. This prosign is also used when you go blank or need some time to think, then you can repeat several <BT> <BT>.

2*.- Although almost all the numbers can be abbreviated it is typical in the RST to only abbreviate the 9. In contests the abbreviated numbers are used more and more although it is unusual to see a single dit being used for a 5.

3*.- Operators with long names usually use a nickname or a shorter version or their name. The same goes with your QTH; a city with a long name can be replaced by a close one with a shorter name.

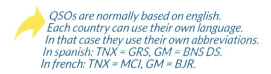

*QSOs are normally based on english.
Each country can use their own language.
In that case they use their own abbreviations.
In spanish: TNX = GRS, GM = BNS DS.
In french: TNX = MCI, GM = BJR.*

❻ We continue and answer

EA2CW

1 - EA7HYD DE EA2CW GM DR OM DAVID =
2 - UR RST 579 57N =
3 - OP MIKEL MIKEL =
4 - QTH BILBAO BILBAO =
5 - HW? EA7HYD DE EA2CW <KN>

Answering. It is an unwritten rule that the QSO can be as long as the calling station (which has the frequency in use) wants it to be.

That is, the answering station should give the same amount of information as the caller. The **QSO should comprise of at least the report and the callsigns. For a typical QSO: callsigns, report, name and QTH. It can be extended with: aerial, equipment, weather, etc**... (see page 50).

The speed at which the contact is made is determined by the calling station, so it is said that:

WHO CALLS, RULES

We continue with the QSO responding more or less as in point 5 of the template, which has already been commented on (see previous point).

❻5.- **HW? EA7HYD DE EA2CW <KN>** we ask how did you receive me (HW?) returning the exchange after sending the callsigns <KN>.

VARIATIONS ❻

❻1A.- **EA7HYD DE EA2CW GM (GD, GA, GE...) DR OM DAVID =** we send greetings according to the time of day, either in our QTH or the location of the station answering or both (GA / GE or GA GE, always yours first). For us it can be night and for our QSO partner, the morning. If we don't know the time at the QTH of the other station we can always say "good day" (GD).

If we want, we can add extra information to the QSO. We will see this later in the "Extending the QSO" section.

❻5A.- **HW CPI? EA7HYD DE EA2CW <KN>** to the how (HW) you can add copy (CPI) to ask, have you received and understood everything correctly?

CW4U.org - **45**

❼ ❽ ❾ Farewell and closing

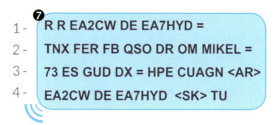

1 - R R EA2CW DE EA7HYD =
2 - TNX FER FB QSO DR OM MIKEL =
3 - 73 ES GUD DX = HPE CUAGN <AR>
4 - EA2CW DE EA7HYD <SK> TU

It is time to finish the QSO. It is best to say goodbye by thanking our QSO partner and signaling the end of the contact <SK>. These are sections 7, 8 and 9 in the QSO template.

❼1.- **R R EA2CW DE EA7HYD =** received / understood (R R). QRA of the two stations and a break (=).

❼2.- **TNX FER FB QSO DR OM MIKEL =** thanks (TNX) for (FER) a good contact (FB QSO) dear (DR) buddy (OM) Mikel followed by a break.

❼3.- **73 ES GUD DX = HOPE CUAGN <AR>** best wishes (73) and (ES) I wish you good contacts (GUD DX), I hope (HOPE) to see you again (CUAGN or CUL) and we indicate that it is going to be the end of this exchange and the QSO <AR>.
Actually in each of the exchanges we could give <AR> but it would be a bit tedious and unnecessary.

❼4 - **EA2CW DE EA7HYD <SK> TU** we send the callsigns for the last time and we end the QSO with <SK> and a "thank you" (TU).

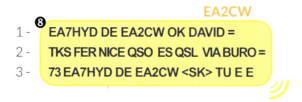

EA2CW
1 - EA7HYD DE EA2CW OK DAVID =
2 - TKS FER NICE QSO ES QSL VIA BURO =
3 - 73 EA7HYD DE EA2CW <SK> TU E E

If our QSO partner knows the **QSO is at an end**, he will finish it following the example in section 8.

❽2.- **TKS FER NICE QSO ES QSL VIA BURO =** thanks (TKS) for (FER) the nice contact (QSO) and (ES) I will send a QSL card via the Bureau (QSL VIA BURO).

❽3.- **73 EA7HYD DE EA2CW <SK> TU E E**   best wishes are sent (73) and callsigns are given again. We conclude by signaling the closing of the QSO with <SK> and the "thank you" (TU), E E. is sent.

It has become customary to finish a QSO with "dit dit", which is like a last, quick goodbye or a "handshake".
It is our QSO partner who will transmit them as his farewell, we should not anticipate them. We will wait to hear the two dits before sending them ourselves or simply finish the QSO without sending them.

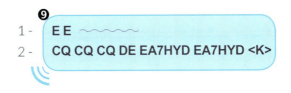

❾1.-   E E   QSO is finished.

❾2.-   We can make a general call once contact has ended:

CQ CQ CQ DE EA7HYD EA7HYD <K>

A radio friend may be listening to the QSO and now wants to make contact with us. He will proceed to send his callsign, AA1ZZZ or EA7HYD DE AA-1ZZZ AA1ZZZ <KN>. Coming back to section 2 for instance.

TU is only used at the end of QSO.

I use a self-built SDR TULIP transceiver and an old restored J38 key.

VARIATIONS ❼❽

❼1.-   **ALL OK DR OM MIKEL =**   there are many ways to say we have received the information. Everything ok (ALL OK, R, VE…). We can also say goodbye to our friend with (OM) and his name.

❼3.-   **88 CU MIKEL <AR>**   in addition to 73 whose sound is very characteristic, at the end of the QSO we can give 88 (best wishes), or 55 (kisses). You choose. QRP lovers use 72 as a "small farewell" to reflect that they are using very low power.

❽2.-   **TKS FER QSO = MY QSL SURE VIA DIRECT =**   I will send you my QSL via direct mail to your postal address (VIA DIRECT).

❽2.-   **TKS FER QSO = ES QSL VIA BURO = PSE UR QSL =**   and (ES) I will send my QSL via the Bureau (VIA BURO) and I ask that you please (PSE) send me your QSL (UR QSL).

❽2.-   **TKS FER QSO = ES QSL VIA LOTW =**   in this case the contact will be validated when the two correspondents upload their logbooks via the internet to the *"Logbook of the World" (http://lotw.arrl.org)*. If the two station logs match, the contact will be deemed to be valid.

❽2.-   **TKS FER QSO = PSE QSL SURE VIA EQSL =**   please (PSE) send a QSL via (VIA) electronic-Internet (EQSL). See *http://www.eQSL.cc* .

❽2.-   **TKS FER QSO = MY DATA IS ON QRZ.COM =**   it is already common to hear stations saying that their data is on (MY DATA IS ON) the QRZ.com website, where sometimes, they include guidance on how to send a QSL.

DOWNLOAD the MP3 file of this QSO at different speeds, site: www.cw4u.org

TRY TO FOLLOW THE DEFAULT ARRANGEMENT OF INFORMATION

● ... extending the QSO, additional information

Among radio amateurs it is normal to talk about the equipment we're using, the antenna, our transmitted power, our hobbies, etc., even the weather in our locality. The latter, more by tradition than by curiosity, since years ago it was difficult to find out, and even interesting to know, the weather in the locality of your contact whereas today it's something we can instantly get from the Internet or TV.

Both stations can exchange as much information as they want. The operator who calls is the one who decides whether to extend the QSO or not. In this case, the information would be between sections 6 and 7 from the template following this order:

RIG - PWR - ANT - WX - TEMP - etc...

❻⁺
1 - R R EA2CW DE EA7HYD = FB MIKEL =
2 - RIG SDR HMBREW ES PWR 10W
3 - ES ANT BAZOOKA =
4 - WX HR SUNNY ES DRY ES TEMP 26C =
5 - HW CPI? EA2CW DE EA7HYD <KN>

The information provided may vary and be more extensive but exchanges are typically like that described in the upper box.

Let's see each part of the message.

❻⁺/ 1.-   R R EA2CW DE EA7HYD = FB MIKEL =   If we start with RR, OK or even C, it means we successfully received the message. (FB) indicates 'good job'.

❻⁺/ 2-3.-   RIG SDR HMBREW ES PWR 10W ES ANT BAZOOKA =
my equipment (RIG) is a self-built SDR (HMBREW or HANDMADE) and my output power is (ES PWR), 10 watts (W), and the aerial is a dipole bazooka (ANT BAZOOKA) (1*).

Weather abbreviations:
SUNNY = sunny   COLD = cold   CLOUDY = cloudy
CLR = clear skies   DRY = dry   FINE = nice
FROST = frozen   FOG = fog   HOT = hot
RAINY = rainy   STORMY = stormy   WINDY = windy
WARM = warm   WET = wet.

Some example of antennas descriptions:

.... ES ANT 3EL YAGI AT 30 M H AGL =   and (ES) my aerial (ANT) is a Yagi (YAGI) with three elements (3EL) at (AT) 30m high (H) above ground level (AGL).

.... ES ANT WIRE 16 M L =   and (ES) my aerial (ANT) is a wire (WIRE) of 16m long (L).
You can use T to send the zeros. That way, 100 W would turn into 1TT W (ES PWR 1TT W). W can also be sent as WATTS.

❻ +/ 4.-   WX HR SUNNY ES DRY ES TEMP 26C =   the weather (WX), here (HR) is sunny (SUNNY) and (ES) dry (DRY) and (ES) the temperature (TEMP) is 26°C (C).

Always make clear that you are using Celsius or Fahrenheit (26°C = 78°F).

❻ +/ 5.-   HW CPI? EA2CW DE EA7HYD <KN>   how do you copy me? (HW CPI?) and we finish with the callsings and closed exchange (<KN>).

EA2CW would give us the weather at his QTH and the details of his station. After the information exchange, we could start a conversation; technical aspects of the station, hobbies, etc.

If you are inexperienced and find yourself in a long QSO, ask for repeats (RPT PSE SRY BUT IM NOVICE) of anything you haven't understood. If you get really stuck try to finish the QSO in a consistent and polite way. Never disappear suddenly.

1- Since the bazooka aerial is a dipole, it is not necessary to say ANT DIPOLE BAZOOKA.*

Kent key

The basics when on-air
you are a newbie but not a "dummy"

Not only must we send a minimum set of information in the right order, we should also do it by following a protocol. Let's remember:

- To learn you must listen, but do so taking into account that you will not always hear things done correctly. You must be clear when mistakes are made.
- By listening you can discover the sending speed of the operator, identify if they are working split frequencies and gather information on them and their station, etc. If you are a rookie, learning this information in advance of a QSO **will make your job much easier**.
- **Never send above your reception speed.** Don't call any station who works at a higher speed than you.
- **It is important not to hesitate during the QSO,** read the example QSOs in the final annex and see how they are performed. At the beginning you can have the sample template on page 77 in front of you.
- In CW you can make contacts even if you don't know a word of the language of the other person. You can do this using common abbreviations, Q codes, prosigns and using the correct order for sending the information.
- **"Who calls, rules".** Remember that the one who starts the QSO should determine the sending speed, the information to be sent and when it ends (short or long QSO).
- **Listen and listen again.** Before calling, check if the frequency is busy (QRL?) and wait. It is possible that there is a QSO here but you cannot hear the station currently sending. However their QSO partner may hear you and they will respond to you with, for example: R, Y, EE, <AR>, QRL, QRX, QSY ... Normally if it is busy, we look for another frequency.
- Make your call (CQ) clearly, indicating your QRA at least twice. If you answer another station calling you also repeat your callsign twice. **Pay attention** in case they ask again. (... PSE RPT QRZ?).
- **Do not extend** your exchange too much. If you are answering send as much information as the other station. Remember, who calls, rules.

European Camelback key - XIX

- Use **EEEEE** or **<BK>** to interrupt the other station during a QSO. The other station will stop and send a question mark (?). This indicates that the exchange has passed to us.

- Except for quick exchanges **<BK>**, you should **always** give the callsigns of both you and your QSO partner at the beginning and end of each exchange.

- Between exchanges leave a **small pause** so that someone can make a quick call and join the QSO.

- If necessary, **shorten** your name or your location. You can say that you are near a known capital. I live in the city of *Chiclana de la Frontera*, but I convey that I am close to Cádiz (NR CADIZ), it is much shorter and easier to send.

- **Be careful** with pile-ups, stations working split, etc (page 70). If you notice something strange in the exchanges or replies, pay attention before calling and listen to that frequency and also the nearby ones.

- **If there is something you don't understand don't panic, just ask.** If you have a mental block simply resume the QSO from the last point you remember.

- If you are using **quick exchanges <BK>**, you must give your callsign every few minutes.

- If you hear a **DX call** or QRZ to a particular zone, respect it and do not answer. CQ JA CQ JA DE EA7HYD EA7HYD <AR> K is an example of a call exclusively for contacts with operators located in Japan.

- Before transmitting, write down the most common information you are going to exchange on a piece of paper (information on your QTH, RIG, PWR, ANT ...), **so you will have them in sight** and you can consult them while you send (use the template in the annex of this book).

- Normally new radio telegraphists operate in the **upper parts of the segment** recommended for CW or around the QRS frequencies. As you go up in speed, you can go down in frequency.

- It is a good idea to "**warm up the fingers**" before sending. Send some letters (without transmitting), to get the feel of the key, adjust the speed, check the equipment, etc.

- Ask for **QRS** (low speed) any time you need it. Don't feel ashamed.

> *Each telegraphist sends with a different and recognizable style, more accentuated with straight or mechanical bug. Thus creating "your signature" or "fist."*

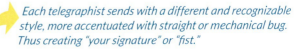

<BK>, quick exchange and its use

If we use <BK> for a long time we should periodically send the callsigns of the stations that are making the contact.

The <BK> prosign is used when we want to interrupt or have a quick exchange within the QSO. For instance, if there is something we missed or couldn't understand, we can send the following:

<BK> PSE RPT UR NAME <BK>   we start and finish the transmission without using the callsigns. We make a quick exchange to ask, please (PSE) repeat (RPT) your (UR) name (NAME).

Our QSO partner will also answer without giving the callsigns:

<BK> NAME DAVID DAVID <BK>   and will pass control of the QSO back to us.

The use of this prosign saves a lot of time in QSOs and facilitates the flow of information.

The <BK> is also widely used in pile-ups. Check out the QSO examples at the end of the book and pay attention to number two.

<BK> is also represented as =

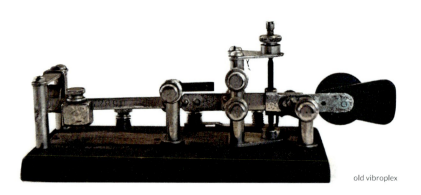
old vibroplex

WHEN IN DOUBT. DO NOT TRANSMIT, LISTEN

TRANSMIT
writing with good "handwriting"

OBJECTIVE: QUALITY AND LEGIBILITY.
Always transmit below your reception speed.

STRAIGHT KEY, PADDLES, BUGS, SIDESWIPERS and **COOTIES** are just some of the names we give the tool with which we 'write' the code.

Important: **Do not practice sending until the sound of the characters and their rhythm are perfectly captured in your mind.**

Remember, if you start sending before you have clearly established the sound of each letter, you will be listening to your own code and this will hinder your learning.

If you don't have a key yet, read this chapter first before buying one and then make a decision.

Keep in mind that to start with, you don't have to have the most expensive or the best key money can buy, you can even build one yourself. There is a vibrant second hand market for keys.

There are many types of key, some more artistic than practical. The following are the most commonly used by radio amateurs worldwide.

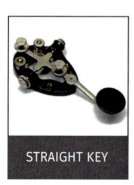

STRAIGHT KEY

This is the most traditional key. The long, straight pivoted arm gives it its name. It derives from the first keys used on the telegraph.

PADDLE KEY

They can have one or two levers that move horizontally. You need an electronic circuit known as a 'keyer' to create the dots and dashes. It's a very comfortable key to use.

BUG

A horizontal, sprung lever creates the dots mechanically but we have to make the dashes ourselves. It needs careful setting up.

The key is something that attracts attention even before knowing the code. It is the tool with which we "write" the messages to be sent and, just as when we learned to write with a pencil and paper, we also have to learn to take the key and use it correctly. Historically most commercial operators had their own key and adjusted it to their own, personal sending characteristics. They never used their partner's key and, of course, they never changed their settings.

It is typical among radio telegraphists to have more than one type or model of key, possibly collected even before knowing how to send and to change the key they use depending on whether they are looking for comfort, speed or simply sending "like we used to".

You must take into account that at the start it can be very difficult to send correctly with a straight key or, indeed, any other type of key. The sounds you produce may be incorrect and **it is very easy to acquire bad habits that can take a long time to unlearn**. Be very careful.

When you start sending, you will do so well below the speed at which you have learned. Do not worry, it is just a matter of time and practice.

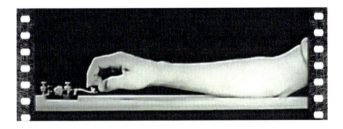

Positioning and handling the key .-

We can contort into a thousand different positions, but the usual setting is to be in our "radio room". In that case we must choose to sit properly in our chair with respect to the table and the key. I will not comment on the possible injuries that we might incur due to sitting badly or having the table at the incorrect height, I'll simply focus on the most common posture of our arm and hand in respect to the key .

You should place your forearm naturally on the table. It is best to place the key about 40 cm from the edge of it so that the elbow is supported, find your position and distance. Some operators transmit with the arm in the air, without support.

 On YouTube you can find videos that really explain the correct posture when it comes to sending. One is from the year 1966: International Morse Code - Hand Sending. *The other is even older, made in 1944. In it we see how to adjust and send with a straight key. It is titled:* US Navy Training Video - Technique Of Hand Sending Morse Code. *The upper image corresponds to the latter* **(see links in the annex).**

Straight key.-

With a straight key **it is up to us to manually form the dots, dashes and spaces and control the length of each.**

How to hold it:

'English or flange' knob, the 'American or button' has no base

- With our arm on the table, we extend it in the direction of the key. We raise our hand to the height of the knob, so that we rest the forearm on the table, almost reaching the elbow, as shown in the picture on the previous page.

- There are quite a few different types of knobs. We usually rest our index finger gently on the top and position the thumb and middle finger either side on the edge of the 'button' or resting on the 'flange'. The muscles of the fingers must allow a light movement and have enough tension to act as a shock-absorber.

- Never hold the key tightly, just let your fingers rest. The little force we make comes from the muscles of the forearm, never the fingers. The arm, hand and wrist are never in tension.

- The main movement comes from the wrist, not the hand. The hand should move up and down freely. The wrist must act as a pivot.

When we start working with any key **we must adjust it to our way of sending** and these settings may change over time as we gain experience. How we set up a key will depend on its design and our personal tastes.

To adjust a straight key:

- Adjust the key spring (spring or magnet) as your wrist demands but note it should not be too tense or slack. To start with you should add extra tension and then gradually slacken it off until you find a setting that is comfortable and that balances the tension and the contact spacing according to your need.

- The initial separation of the contacts should be about 1.5mm and this will support slow speed sending. As we improve, we will close this distance to 0.75 mm or even less for more speed.

- Another adjustment that we can make is the pressure on each side of the axis on which the arm pivots. We must be careful because with this we can make slight horizontal adjustments to the lever and move the contacts, which should always be totally flush and aligned to each other.

- We should never feel a vibration or a bounce in the lever.

Single paddles, Sideswipers and twin paddles.-

Single and twin paddles require an electronic circuit called a 'keyer' that generates the dots and dashes with the correct lengths and inter-symbol spacing (like a machine gun for dots or dashes). This device allows us to control the sending speed and can be included within our transceiver, built or bought separately. Normally it has memories to, for example, be able to automatically record and send something that we repeat a lot, such as our callsign.

This is how we use it:

- With a **single lever paddle** we typically make the dits by pressing the paddle to the right with the thumb and the dahs by pressing the paddle to the left with the index finger (for right handed operators). In the keyer we can swap this around if we prefer.

Kent key single paddle

- Support your arm on the table in the same way as described with the straight key and relax the wrist so that we can swing the hand slightly to the sides and so touch the paddle with the thumb and finger.

- With a **twin level paddle** you normally make the dots using the left paddle and dashes using the right, but it would be a waste to use it as if it were a single level paddle. **The normal thing with a twin lever paddle is to make the characters by squeezing or pressing the two levers at the same time.**

With iambic keys, **both paddles are used but in a different order, that way we obtain all letters with only two moves**. It is a fast and effective method -transmit faster and with perfect characters-. We can find more detailed information on the following page (*see also annex, videos*).

 The ancient iambic Greek poetry has a metric, a rhythm, formed by a short and a long syllable.

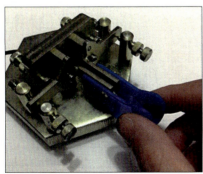

how to hold a twin paddle, key by EA4GKY

There are two modes of iambic sending:

- Mode A: complete the character that is being sent when the paddles are released.

- Mode B: when the paddles are released, the keyer will complete the character being sent with an element (dot or dash) opposite to the last one sent.

Example: to send the letter C using the mode B, we press both paddles (DAH before DIT) and let them go after listening to the last DASH and the keyer will add the remaining DIT; in mode A, we don't let the paddle go until we hear the second DIT.

You should handle the iambic key gently, giving small touches to the paddles. Take it easy, there's no need to slap it about, that would be tiring and painful for both of you (HI).

Adjusting a paddle key:

Note that we are actually adjusting two horizontally mounted straight keys. Each model will have its peculiarities, but as with straight keys, the main thing is to control the separation between contacts and the tension or force in the springs or magnets. We usually aim to make the same adjustments on both paddles.

twin paddles

Managing an iambic key:

Start by setting the keyer at a low speed, about 15 wpm, and try sending dits and then dahs. Then alternately press the paddles, release them separately and listen to what happens. Now set the keyer to use **mode B**, which is the **most popular mode**.

IAMBIC MODE B – PADDLE TAPS					
A	L, R	N	R, L	www.cw4u.org	
B	R, L	O	R		
C	R, L	P	L, R, L		
D	R, L	Q	R, L	1	L, R
E	L	R	L, R	2	L, R
F	L, R	S	L	3	L, R
G	R, L	T	R	4	L, R
H	L	U	L, R	5	L
I	L	V	L, R	6	R, L
J	L, R	W	L, R	7	R, L
K	R, L	X	R, L, R	8	R, L
L	L, R	Y	R, L	9	R, L
M	R	Z	R, L	0	R

Try sending a letter 'A' every second. **If you already know how to send an A, try not to release the left paddle (L) and then release the right one (R) you will get a letter R or an L depending on how quickly you release the left paddle.** You will have pressed this sequence for the R: L, R and release both, first the right and then the left. This is what the iambic code generator is responsible for.

Here are a few more examples:

- **Letter B,** the sequence would be: R, L until we have heard the three DITS and then let go.

- **Letter C**, is a little more complicated. Keep the R paddle pressed and then immediately press the L. Press and hold R until after the second half of the DAH and then release it and end by releasing the L at the second DIT (mode A) or during the second DAH (mode B). **As you can see there are only two touches and not four.**

In the table on the previous page, **almost all the letters are made with just two paddle touches in iambic mode B**, which speeds up the sending and allows us to create messages much faster than with a straight key, in addition to setting the pace with more quality. Remember, when you are learning this system, you should never count DITS or DAHS.

Continue practicing little by little with each letter in iambic mode B until you master it. Then you can try mode A and choose your favorite. Consider the characteristics we saw of each of the iambic modes.

At first it may seem difficult, but once you get used to it, you will succeed. This doesn't mean that you shouldn't also practice and make contacts with straight keys or any other type.

Mechanical Bugs, Vibroplex.-

It may be the most singular key of all and that is why it is used by experienced radio telegraphists. It is certainly not a key for beginners. Vibroplex is a brand that has been manufacturing them for years.

It generates the dots mechanically, but we make the dashes. Its adjustment of springs, contacts and counterweights is delicate and, when we use it to send, it has a very particular sound. If you are curious, I recommend you search online for a video where you can see how it is used.

Starting to "tap"
sending is different to receiving

The process of sending is totally different from that of receiving. **Your brain must learn a new way of 'writing'**, so it is important to have heard a lot of code before starting to send.

We have already seen that it is important to learn what good code sounds like, so the ideal is to learn all the characters before starting to send. You are going to be keen to start sending but for now I don't recommend it. If you are not completely familiar with the sound of each character, its timings, its rhythm, then you won't be able to send them correctly. If you must start learning to send then it's much better if you can find someone with experience to supervise you and correct any errors at the outset.

TRANSMIT WITH ACCURACY AND QUALITY

As we have seen we can use many different types of key, each with its own characteristics. **With a straight key it is you who determines the timing and rhythm and if you use a paddle, the electronic keyer will do it for you**. This latter approach is much more interesting **when learning to send Morse code** because we automatically and impeccably generate the correct duration and spacing of the characters in addition to getting used to the correct sound. Its use seems more complicated and requires a little more practice than the straight key, but once you get used to it, your code will gain in speed and quality.

Choose whichever type of key you like to start and do not be discouraged. To handle any type of key correctly takes time.

It is good practice, before we start sending, to **fix the key to the table** to prevent it from moving. If you operate portable you can attach it to your leg or a board where you can also place your log book, etc. When in your shack you should fit a weight to the base of the key or use some type of removable adhesive. Blu Tack or the type of plasticine used to hang pictures is perfect since, in addition to fixing the key to the table, it can easily be removed and replaced without losing adhesion (or staining).

NEVER SEND FASTER THAN YOU CAN RECEIVE

IF YOU MAKE A MISTAKE WHEN SENDING, CORRECT IT AND CARRY ON

When you send, especially at the beginning, you may suffer with nerves and make a mistake. If that happens you must point it out with several E's or three well separated E's in a row (E E E), then repeat the problem word and just forget about the mistake. **If you keep thinking about it you won't be able to concentrate on the rest of the QSO.**

Connecting the key to the transceiver.-

Most commercial or self-built transceivers will have a socket to plug in a key and the terminals of your key will need to be connected to an appropriate plug (see the manual for your radio).

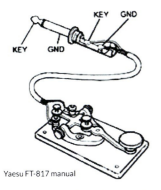

Yaesu FT-817 manual

There are two types of cables, one for straight keys with mono plug (2 wires) and another for paddles with a stereo plug (3 wires, or 2 wires and ground). Normally our station will have an internal electronic keyer. If not, we must build or buy one. In that case the paddle key will be connected to the external keyer and then this will be connected to your transceiver, usually with plug connections as shown in the images.

Note that if you use an External keyer you will connect this to your transceiver as if it were a straight key and put the transceiver in 'straight key mode'. This is because the keyer is making the DITS and DAHs so the radio doesn't have to.

Use a shielded cable (better if flexible) long enough to reach the transceiver and a little more in case one day you change the key or radio. Always be sure to make good soldered joints on the plug.

Dots and dashes generator.-

If your transceiver has an internal keyer, check the manual and look at the plug connections before soldering. If you don't have a built-in keyer and you want to use paddles you will have to search the Internet for a circuit, a kit or buy a commercial one. There are many models and prices, some in addition to generating the dots and dashes have other very useful options such as memories. Check also if it has both iambic modes of sending (A and B). I recommend you look for one that is easy to use and setup.

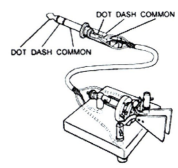

Practice with an external oscillator or sidetone.-

Some operators practice with the sidetone of the radio without actually transmitting, just listening. If you don't want to do this, you can practice with a straight key using a simple oscillator or "buzzer" and a battery or, if it is with a paddle, you will need to use an electronic keyer. As I said, on the Internet you will find many ideas and even circuits already assembled.

Monitor and correct your code.-

At this time (and only at this time) you can enlist the help of devices, programs or applications for phones that decode the Morse code. In this way you can **check the quality of your code** and know if you are correctly keying the dots, dashes and the spaces. It is a very useful aid, especially at the beginning.

Free Android app:
MORSE CODE READER

The drawing above is one of the first keys used in Morse demonstrations. The one on the right corresponds to the key designed by Vail, called the "Correspondent".

Key maintenance.-

With time and humidity the contacts of your key can get dirty so clean them gently. Never pass an abrasive sandpaper between them, especially if they are gold or silver plated. You can simply slide a sheet of paper between the contacts while gently closing them (without squeezing too much) to remove dirt and oxidization.

Pay attention to moving parts and bearings (if any). Occasionally check the key contacts. It wouldn't be great to be disconnected in the middle of a QSO because the cable is in poor condition.

Before transmitting check that everything is in place and working.

SPACING IS AS IMPORTANT AS THE DOTS AND DASHES

PROCEDURES
to keep in mind

Only the basic information will be provided here. Further information can be found on the internet or in books. First of all, we have to know that we can transmit CW in the following ways but some of them are not available in all transceivers:

- **SIMPLEX**: we transmit and receive on the same frequency (zero beat). This is the most common mode.

- **DUPLEX**: we transmit on one frequency and receive on another. For example: pile-ups, DX.

When available we must also configure our transceiver (*) to determine **the delay between transmission and reception** when we use semi break-in mode. This can be short or extremely short or you can even listen between dots and dashes (full break-in). We have the following options:

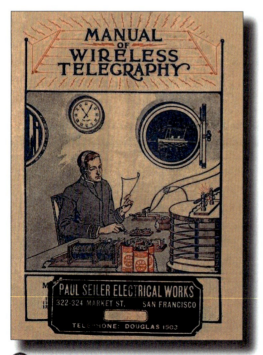

Break-in.-
We only listen after we stop transmitting.

Semi break-in.-
We set a "delay time" so that we can still receive between the words of our transmission.

Full break-in.-
Like semi break-in but now we can listen between the dots and dashes of our own transmission.

*In transceivers with older components or designs, very rapid changes between transmit and receive may not be possible due to switching problems.

> On the subject of pile-ups, working DX, etc... You can read much more in the book "Ethics and Operating Procedures for the Radio Amateur", see annex.

Specific CQ calls
searching for a particular contact

You won't only hear the normal call **CQ CQ CQ DE EA7HYD EA7HYD <K>**. There are more ways to call, depending on our goals.

For example:

CQ DX CQ DX DE EA7HYD EA7HYD CQ DX <K>
With this call we make it clear that we want to make long distance contacts.

CQ NA CQ NA DE EA7HYD PSE <K>
We make a call to a specific area, in this case to North America (NA).

Other areas could be: EU = Europe, AS = Asia, OC = Oceania, AF = Africa, SA = South America, JA = Japan...

CQ CQ DE EA7HYD PSE ONLY EU <K>
we make a call only looking for contacts within Europe.

CQ EA1 CQ EA1 DE EA7HYD <K>
we call a specific region (EA1). For example, in a contest you may be missing that region.

It is normal to hear the CQ sent as a **single sound** in which the two letters have been joined together. Another type of call is that used for **contests** which we will see on the next page.

Begali Magnum

Sorting the answers to our call.

Sometimes more than one station answers and you have to decide which to respond to first: the one that has the callsign with the prefix that interests you, the one that has a tone that you find easiest to hear (higher or lower pitched), the one that is sent with clarity and quality or simply according to the order you hear them...you must choose.

Check out the pile-up section (page 70) to learn how to deal with more than one station at a time. This is a topic that, no matter how much you read, you will not understand in depth until you put it into practice. Pay close attention to the way stations work pile-ups and also to those calling it. It will be the best way to learn.

Contacts in a contest
organized chaos

 Every contest has its own set of RULES which are easy to find on the internet. They will indicate: the purpose of the contest, the mode of transmission (CW), the bands, power to use, duration: dates and times in which to operate, the call (CQ TEST or other), how to get points, multipliers, the exchange to be made and where to send the list of contacts made. In each contest there are different categories that participants can enter under depending on things like the number of operators, number of radios, power or bands or a combination of these, etc...

One of the most common activities in CW, especially at weekends, are contests. They test the skill of radio amateurs but, when you are a rookie and listen to a contest it's like you are hearing wild animals sending dots and dashes at full speed and making no sense. This can be intimidating and make you wonder if you can ever take part.

Well, I say yes you can. There are many different types of contest, some of them are even targeted at beginners. You may not win a prize but at least you can participate and become accustomed to them.

> **IN A CONTEST QSO, ONLY CALLSIGNS AND EXCHANGES ARE SENT. NOTHING ELSE**

Each competition has its peculiarities and if we don't know them and start making contacts we are likely to screw up and annoy people who want to win awards. More than at any other time, in competitions, first of all... LISTEN.

Contest categories:

- **SOAB LOW:** *(Single Operator All Bands Low Power)* One operator, all bands, low power (up to 100w).
- **SOAB HIGH:** *(Single Operator All Bands High Power)*, more than 100w.
- **SOAB QRP:** *(Single Operator All Bands QRP)* QRP up to 5w.

  These categories can also appear on MONOBAND (only one band):

- **M/S:** *Multi-operator, one transmitter.*
- **M/M:** *Multi-operator, Multi-transmitter.*

  Multi-operator categories are typically AB (all bands) and sometimes there are LOW / HIGH categories with the same power criteria used in the Single Operator categories.

> # A contest QSO is quite unlike an ordinary QSO, here BREVITY is essential

The main goal is to make the maximum number of contacts on a specific date and time, on specific bands and under certain technical conditions.

Each contact is worth a number of points which are multiplied depending on the number of countries worked, distance or conditions determined by the rules. So **what matters most is to make lots of contacts, but without losing sight of the multipliers**. In this way if you make lots of contacts but not many multipliers, your score will not be great and, similarly, if you make only a few contacts but with many multipliers, the result will not be good either. Planning a strategy according to your working conditions will be very useful. In some cases, several radio amateurs group together to form a team to compete.

After the contest you normally need to email a **list of the contacts** (log) made along with the data requested in the rules in a specific format (Cabrillo). The event organizers collect the data and publish a list of the corrected scores. Normally we use a computer and a specific contest logging program to log each QSO in real time so that, when we finish the contest, we already have a complete log with the points and multipliers achieved already calculated.

Every contest has its own organizing committee who decide whether someone has not complied with the rules and apply penalties if necessary.

IMPORTANT: in a contest QSO only the callsign, report and some specific information is exchanged e.g. serial number, province, age...

Example of a contest QSO:

On the left we have an example in which the station EA2CW is calling CQ TEST, it is answered by EA7HYD and then by EA3GZA. As you can see, we do not exchange greetings, prosigns or any unnecessary information. We only give the callsign, report and data (**BI**, **CA** or **T**, the license plates of the provinces, in this case).

One of the bits of information to pass in a contest is the RST. It is common practice, for speed and to avoid errors, for everyone to send 599, not the real report -it's a contest thing!-

Working split mode, pile-ups...
a QSO that is not what it appears

If you hear a station calling in a concise way, then answering with a callsign and report and then quickly repeating the process between empty spaces, without anyone appearing to answer, don't transmit. Don't call the station here but, instead, look a few kilohertz above or below the frequency. It is likely you will find several operators calling or giving their callsign and report one after another. You just found a group of radio amateurs working in a very particular way.

It's quite normal when a station operates from an unusual location that a lot of stations want to try and contact them and suddenly the frequency can become very busy as all the stations step on each other trying to call the DX. This tremendous fuss is called a PILE-UP.

The operator or operators of the calling station (DX expedition, unusual country, special award or special event station) are choosing which stations they answer by changing their receiving frequency. Therefore the contact is made in an odd way, the DX station is calling on one frequency but receiving on another. This is called working SPLIT.

If you want to avoid making a mistake by calling where you shouldn't, you need to find out where the station is listening and the best way or method to make contact. I advise you to do the following:

> **LISTEN AND LISTEN AGAIN,**
> **BE PATIENT**

How to have a split-mode QSO:
- Before launching into sending your callsign to a station that is saturated with calls, listen for a while to see how the rest of the operators are working. In this way you can discover the approach they take and adapt to the rhythm and speed of operating.
- From time to time you will hear the DX station send 2 UP (for example). This tells us that it is listening up to 2 kHz above the frequency on which they are transmitting. Alternatively 2 DN indicates that they are listening up to 2 kHz below. If a figure is not indicated (UP), the normal is ~ 1 kHz.
- Tune into the signal of the DX station and then use the Split function on your transceiver to set your transmit frequency where you think the DX is listening.
- Send your callsign ONLY once per call. Listen for an opportunity, a gap to enter and, when you call, just follow the QSO format of the previous operators.
- NEVER indicate that you are working /QRP, etc., avoid sending unnecessary information in a pile-up. NEVER put DE in front of your callsign as it can be confusing and of course, NEVER repeat the callsign of the called station (DX).

When working SPLIT, QSOs are typically very short, contest type exchanges (callsign, report and thanks) passing immediately to the next station. Below is an example of a DX station working split. It's a little artificial but it serves to illustrate split operation. The two working frequencies (TX and RX) of the DX station are represented in separate columns. The broadcasts of the DX station (P2LMN) are in **red** and his reception in **black**. An explanation of each step and key examples are in **blue**.

TX	RX	
7.008	7.009 7.010 Mhz	*The DX station controlling the PILE-UP transmits on 7.008 MHz and receives the calls around ±7.010 MHz (2 KHz above).*
	JA4FGH	*The stations make their calls and send reports on ± 7.010 MHz, the frequency where DX is listening.*
W8KVN 5NN		*The DX station gives control to W8KVN.*
	EA2CNA	
	5NN TU (W8KVN)	*W8KVN sends a report and thanks to P2LMN (is for clarity, not transmitted).*
	ZN5HJK	
AA8DFG/QRP		*The station AA8DFG calls on an inappropriate frequency and adds a suffix to their callsign /QRP, which is not correct.*
TU P2LMN UP 2		
	JA4FGH	
JA4FGH 5NN		
	AA8DFG/QRP	*The DX station thanks W8KVN and indicates that it is receiving 2KHz above (for "the clueless").*
	5NN TU (JA4FGH)	
TU P2LMN UP	AA8DFG/QRP	
	EA2CNA	*AA8DFG calls again, this time on the correct frequency, but still adding /QRP to their callsign and also calling out of turn.*
EA2CN? 5NN		
	EA2CNA 5NN TU	
EA2CNA TU		*AA8DFG calls again immediately which is unnecessary and still at the wrong time.*
	ZN5HJK	
ZN5HJK 5NN		*The rest of the stations call and answer as the DX station calls them.*
	5NN TU (ZN5HJK)	
TU UP2		
	AA8DFG	*You need to look for a gap to enter into the pile-up and give your callsign only once.*
	DL4WER	
AA8DFG 5NN		*Control is given to AA8DFG.*
	AA8DFG/QRP 599 TU HPE CUANG = GL ES GDX <SK>	*AA8 answers, but incorrectly, providing unnecessary information in a pile-up.*
AA8DFG NIL DL4WER 5NN TU		*DX station tells AA8DFG off (NIL) and gives control to DL4WER.*
	5NN TU (DL4WER)	
TU P2LMN UP 2		*The pile-up continues.*
.... keep going		

Miscellany
In case of doubt, ask

- First and foremost the most important thing to do is: **LISTEN**.

Listening will avoid problems and misunderstandings. Be patient and listen, it's the best way to learn. Historically, in many countries, it was compulsory to listen for a few months before getting a license.

- We all suffer from mental blocks. If it happens, don't stay silent, just explain that you are inexperienced (**SRI BUT HR NEW OP PSE RPT**). Don't do as may do and say you have QRM, QRN, QSB... be honest

- Sometimes operators answer **R R** to the message they just received, but they don't answer the questions they've been asked suggesting that they didn't receive everything. If this happens to you it is better to ask for repeat of the information.

- If there is something you don't understand, ask for a repeat (for example **PSE RPT NAME, PSE AGN UR NAME, NAME AGN**) or, if the Morse is too quick and you need the other operator to slow down send **PSE QRS**. Remember you can use **<BK>** for quick exchanges.

- The letter R is also used as a comma between numbers so 12,35 would be sent as **12R35**. Similarly you can use R to separate hours and minutes like 14R30H, but you can also simply send **1430H**.

- In some languages the letter Ñ is used which can be sent like **NH** or **NN**.

- When sending a mix of whole numbers and fractions separate the whole number and the fraction with a dash. Example: 1 ¾ send 1-3/4. For ¾ 8 we would send 3/4-8.

- If you want to send a number followed by the percentage, example: 3%, send 3-0/0 and not 30/0, you must include a hyphen to separate the number.

- Many radio hams do not indicate the make of their transceiver (**RIG**) and only give the power (**PWR**) and aerial (**ANT**), thereby avoiding the advertising.

- Difficult words are sometimes sent with more spacing between letters to make it easier to read (cities names, etc...)

- When in the QSO we want to say good morning, good evening or good night but we don't know what time it is in the other operator's country, we can send **GD** (good day), it may not be 100% correct, but at least we're saying hello.

unknown

- In addition to sending eight dots or three separated (**E E E**) to indicate an error, some operators send a question mark (**?**). After this notification we resend the whole word.

- If you want to make a contact on a frequency recommended for **QRS**, but it is in use by a station transmitting **QRQ** simply change frequency up or down a bit. **QSY** and everything will be sorted... nobody owns a frequency.

• Don't get obsessed with the speed. We don't need to transmit fast (QRQ). We don't get paid as commercial radio telegraphists to send fast telegrams. It's about enjoying the hobby with like-minded people. Call at a speed you feel comfortable with and someone will reply.

• At first it is common to make mistakes... don't worry about it. We all make mistakes, even great telegraphists make them. **What is important is that you send with clarity and quality.**

• You must pay attention when tuning the radio to the frequency of the station you want to contact. This process is called **zero beating** or **netting**. Some radios do this automatically, otherwise tune to the other station by ear until the pitch of your sidetone matches the received signal.

• Historically, on valve sets, it was quite normal for our **tone** to change as we transmitted. The report **K** was used to indicate our signal had key clicks and **C** was used to report 'chirp' on our tone.

• A common sidetone frequency is 800 Hz, but there are operators who prefer 700 Hz or as low as 400 Hz. Experiment and see what works for you.

• Personally I prefer to take notes with a **pencil or propelling pencil**. Pens can leak at high temperatures and stop working at low temperatures and don't work well on damp paper.

• You can use the **RIT** (Receiver Incremental Tuning) or clarifier to make slight changes to the tone of the received signal by adjusting the received frequency without changing the transmit frequency. Don't forget to switch it off when changing frequency.

unknown

• When receiving, play with the **filters** to adjust the bandwidth to the value you need to isolate the tone of a station from adjacent stations or QRM.

• Create a **QSO template with the most common information** you wish to send, your city, equipment, etc. Start with the example QSO template in the Annex (p. 77). Study and memorize the structure and then practice off-air until you can send the information without any problems. Memorizing the basic QSO is very helpful and gradually you can build on it.

• They're rare and we probably shouldn't use them too much, but we also have insults in CW. Especially for those who don't meet the expected standards!

They include:

NIL.- "Not in log", something about the QSO or the other station was so bad you do not log the contact. Be careful, it also means I HEAR NOTHING.

99 .- Something negative, the same way we use 73, 88 or 55 as something positive. You can also send this as **NN**.

LID .- You are a bad operator.

REMEMBER:
THERE IS ALWAYS SOMEONE LISTENING

THE BEGINNER'S NIGHTMARE

Morse Code Table

A	.-	1	.----		MORSE CODE Table
B	-...	2	..---		
C	-.-.	3	...--		.-.-.-
D	-..	4	....-	,	--..--
E	.	5	.....	:	---...
F	..-.	6	-....	;	-.-.-.
G	--.	7	--...		
H	....	8	---..	!	-.-.--
I	..	9	----.	'	.----.
J	.---	0	-----		
K	-.-			(	-.--.
L	.-..			)	-.--.-
M	--	Ä	.-.-	"	.-..-.
N	-.	Á	.--.-		
O	---	Å	.--.-	?	..--..
P	.--.	Æ	.-.-		
Q	--.-	Ch	----		
R	.-.	Ç	-.-..	+	.-.-.
S	...	É	..-..	-	-....-
T	-	Ï	-..--	×	-..-
U	..-	Ñ	--.--	/	-..-.
V	...-	Ö	---.	=	-...-
W	.--	Ô	---.		
X	-..-	Ø	---.		
Y	-.--	Ü	..--	@	.--.-.
Z	--..	Ź	--..-.	$	...-..-

Here you have a table with the representation including dots and dashes of every character. But, **remember Morse code is made of sounds, not symbols,** and if you want to learn it correctly, you should never visually memorize the representation of the characters. Symbols and some particular letters of some languages are included (there are many more).

Remember, if you want to learn the Morse code fluently, when you practice you should never have this table or a similar one in front of you.

Your first QSO template

Fill the empty spaces and you will have a template to start making contacts, **you can modify** it to your liking. The boxes with continuous line borders are **IF YOU CALL** and the boxes with dotted borders are **IF YOU ANSWER**.

QRL? (wait a few seconds) QRL? (wait again)

CQ CQ CQ DE _____ _____ PSE <K>

_____ DE .... <KN>

.... DE _____ GD DR OM ES TNX FER CALL =

UR RST .... =

NAME 1.................   1................. =

QTH .... =

HW? .... DE _____ <KN>

_____ DE .... GD DR OM 1.................. ES TNX FER CALL =

UR RST .... =

NAME 2...................   2................... =

QTH .... =

HW? _____ DE .... <KN>

R R .... DE _____ =

TNX FER FB QSO DR OM 2........................... =

HPE CUAGN 73 <AR> .... DE _____ TU <SK>

_____ DE .... OK =

TNX FER QSO = 73 ES DX

_____ DE .... <SK>

CW4U.org - 77

Quiz
curiosities, history

HOW DID MORSE GET THE IDEA?

In 1825, the established painter Samuel Morse was in Washington D.C. making a portrait of the Marquis de La Fayette, French hero of the war of independence, when a messenger came at a gallop to inform him that his wife had fallen ill. The next day he received another letter from his father announcing the sudden death of his wife. Morse did not finish the portrait and travelled as fast as he could to his home.

He arrived too late, his wife had been buried. He decided to devote himself to the search for better methods for long-distance communication. It was the origin of the telegraph.

The future inventor studied Religious Philosophy, Math and Equine Veterinary in Yale College. He studied electricity with Benjamin Silliman and Jeremiah Day. During his student years he discovered painting and became one of the most important portrait painters in the country.

He tried to introduce the telegraph in Europe and the United States but it only succeeded when Congress approved a project to build a telegraphic line between Baltimore and Washington.

Morse was involved in long litigation over the rights of the invention with other scientists and countries, since the same invention was being developed at the same time in other places.

He died at age 80 after winning a great fortune.

WHAT WAS THE INVENTION OF THE TELEGRAPH BASED ON?

Morse developed the concept of a single wire telegraph after seeing several experiments with electromagnets and relays made by **Joseph Henry**, one of the discoverers of the electromagnetic induction, but it took so long to publish it that the discovery was credited to **Michael Faraday**.

In 1831, J. Henry invented the telegraph and in 1835 perfected it so that it could be used over long distances. However, he did not patent it. It was Samuel Morse who, personally helped by J. Henry, put into practice the first telegraph in 1839 between Baltimore and Washington.

He devoted his life to fostering the development of new sciences and the exchange of scientific ideas worldwide.

WHO DEVELOPED THE CODE?

Morse did not invent the code that bears his name. The idea of encoding each character as a sequence of dots and dashes and defining its measurements over time can be attributed to Alfred Vail. Work he did while collaborating with Morse in the creation of the electric telegraph. Vail also designed the first telegraphic key called the "Lever Correspondent".

He met Morse in 1837 when he attended one of his first experiments as a spectator. Vail helped Morse refine his prototype and they were partners for years, but he retired from the business in 1848 because he believed that Morse's line managers did not value their contributions sufficiently.

A PAINTER'S PROTOTYPE

By 1835 Morse had created his first telegraph model. The apparatus was operated by means of a crank that moved the "sign" to be transmitted. The signs were generated by drawing metal dies like a comb with the teeth representing the dots and dashes under a pivoted wooden arm. This switched an electrical circuit and activated a solenoid (invented not long before) in the receiver which in turn printed the marks onto a paper tape.

Below is a picture showing a model of one the "signs" that were used to open or close the circuit.

Morse built his first design in a painter's canvass-stretching frame. In the upper central part of this frame hung a pendulum at the end of which was a pencil. Each time the circuit was closed, an electromagnet would draw the pendulum towards it, when the circuit was opened the pendulum would return to its original position. The pencil reproduced these movements on a paper tape that was moved by rollers and gears.

After this process, a zig-zag line was visible on the paper, which reproduced the notches of the teeth and slots in the comb of the transmitter. The recorded signals were translated into the corresponding signs. Initially, Morse's idea was to transmit figures that would later be "translated" into words.

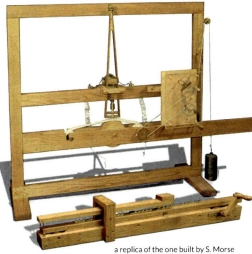

a replica of the one built by S. Morse

WHAT WAS THE FIRST MESSAGE SENT?

The first effective and complete transmission that was made with the new system was on January 6, 1838 at the Speedwell foundry owned by Alfred Vail's father, using about 3 km of telegraphic line. The message read "A patient waiter is no loser".

The first official telegraphic message was sent on August 24, 1844 and read: "What hath God wrought", A biblical quote chosen by Annie Ellsworth of Lafayette.

NEWS FROM THE FIRST MOMENT

All the newspapers of the time reported on the invention including, of course, the Scientific American magazine that launched in August 1845 and talked about the invention in its first issue:

"This wonder of the age, which has for several months past been in operation between Washington and Baltimore, appears likely to come into general use through the length and breadth of our land.", "...it is contemplated by the merchants of our Western states, to communicate their orders for goods, etc. by means of the telegraph, instead of abiding the slow and tedious progress of rail-road cars."

FIRST USE OF SOS, CQD

The first time SOS (Save Our Souls) was used was on 10 June, 1909 from the passenger ship RMS Slavonia which ran aground and sank near the Azores.

CQD was the distress signal used in telegraphic transmissions at the beginning of the 20th century. CQD was interpreted as "Come Quickly, Distress", but the real meaning is: CQ ("Copy Quality", general calling code) and D ("Distress").

Telegraphs had traditionally used "copy quality" (CQ) to identify messages of interest to all the stations along a telegraph line and this is still used today by amateur radio operators to initiate contact (CQ sounds like seek you). The prefix CQ was transferred to the radio and ships, but neither the telegraph nor the radio had a distress call, so Marconi suggested using a "D" to highlight that there were problems.

HELP MESSAGE FROM TITANIC

"CQD CQD CQD CQD CQD CQD de MGY MGY MGY MGY MGY position 41.44 N 50.24 W". The callsign of the Titanic was MGY. With the use of voice the SOS message changed to MAYDAY.

GLASS ARM

Now known as "carpal tunnel syndrome", it was a painful condition in the tendons of the wrist suffered by telegraphists who mainly used straight keys. It was caused by poor posture and overwork over a protracted period.

SOLAR STORM, 1859

The Smithsonian magazine reported in the summer of 1859, that tremendous solar eruptions had destroyed the telegraph line between North America and Europe.

During the storm the Northern Lights (aurora borealis) could be seen in the south of Mexico and the Caribbean Sea.

INVENTOR'S MARRIAGE PROPOSAL

Thomas Alva Edison worked as a telegraphist since he was 15 years old. Some years later he taught his girlfriend and future second wife, Morse code so they could communicate secretly in front of their parents. One of the messages asked, in dots and dashes, "will you marry me". She responded in the same way, "yes".

SPEAKING LIMITATIONS

Morse code has been used as a conversation aid for people with a wide variety of communication disabilities.

OTHER WAYS TO SEND THE CODE

Morse code is not only sent via wire, radio or sounds (horn), it can also be sent with light. You may have seen it used between Naval ships in the movies.

"YOU PULL HIS TAIL IN NEW YORK AND HIS HEAD IS MEOWING IN LOS ANGELES." Albert Einstein

"The wireless telegraph is not difficult to understand.
The ordinary telegraph is like a very long cat.
You pull the tail in New York, and it meows in Los Angeles.
The wireless is the same, only without the cat."

LUDWIG VAN BEETHOVEN

In the opening of his fifth symphony Beethoven repeats the letter V several times. This part of the symphony was broadcast on radio programs by the Allies in World War II, where "V" was for Victory.

WHEN MORSE DIED IN 1872

More than 650,000 miles of telegraph wire connected cities around the world. At the beginning of the 20th century, messages started to be sent wirelessly via radio.

Abbreviations

- AA All after (used after question mark to request a repetition)
- AB All before (similarly)
- ARRL American Radio Relay League
- ABT About
- ADR Address
- AGN Again
- ANR Another
- ANT Antenna
- ARND Around
- AS Wait
- BCI Broadcast interference
- BCNU Be seeing you
- BK Break (to pause transmission of a message, say)
- BN All between
- BTR Better
- BTU Back to you
- BUG Semiautomatic mechanical key
- BURO Bureau ("Please send QSL card via my local/national QSL bureau")
- B4 Before
- C Yes; correct
- CBA Callbook address
- CFM Confirm
- CK Check
- CL Clear (I am closing my station)
- CLG . . . Calling
- CONDX Conditions
- COS Because
- CQ Calling ... (calling all stations, any station)
- CS Callsign
- CTL Control
- CUAGN See you again
- CUD Could
- CUL See you later
- CUZ Because
- CW Continuous wave
- CX Conditions
- DE From (or "this is")
- DN Down
- DR Dear
- DX Distance (sometimes refers to long distance contact), foreign countries
- EMRG Emergency
- ENUF Enough
- ES And
- FB Fine business (Analogous to "good")
- FER For
- FM From
- FREQ Frequency
- FWD Forward
- GA Good afternoon or Go ahead (depending on context)
- GE Good evening
- GG Going
- GL Good luck
- GM Good morning
- GN Good night
- GND Ground (ground potential)
- GUD Good
- GX Ground
- HEE Humour intended or laughter - often repeated twice i.e. HEE HEE
- HI Humour intended or laughter
- HNY Happy new year
- HR Here, hear
- HV Have
- HW How; How copy
- II I say again
- IMP Impedance
- K Over
- KN Over; only the station named should respond (e.g. EA2CW DE EA7HYD KN)
- LID Poor operator
- MH Meters high (antenna height)
- MILS Milliamperes
- MNI Many
- MSG Message
- N No; nine
- NIL Nothing
- NR Number; Near
- NW Now
- NX Noise, noisy
- OB Old boy
- OC Old chap
- OK Okay
- OM Old man (any male amateur radio operator is an OM regardless of age)
- OO Official observer
- OP Operator
- OT Old timer
- PLS Please
- PSE Please
- PWR Power
- PX Prefix
- R Are; received as transmitted (origin of "Roger"), or decimal point (depending on context)
- RCVR Receiver
- RFI Radio-frequency interference
- RIG Radio apparatus
- RPT Repeat or report (depending on context)
- RPRT Report
- RST Signal report format (Readability-Signal Strength-Tone)
- RTTY Radioteletype
- RX Receiver, radio
- SAE Self-addressed envelope
- SASE Self-addressed, stamped envelope
- SED Said
- SEZ Says
- SFR So far (proword)
- SIG Signal or signature
- SIGS Signals
- SK Out (prosign), end of contact
- SK Silent Key (a deceased radio amateur)
- SKED Schedule
- SN Soon
- SNR Signal-to-noise ratio
- SRI Sorry
- SSB Single sideband
- STN Station
- T Zero
- TEMP Temperature
- TFC Traffic
- TKS Thanks
- TMW Tomorrow
- TNX Thanks
- TRE There
- TT That
- TU Thank you
- TVI Television interference
- TX Transmit, transmitter
- TXRX Transceiver, transmitter + receiver
- TXT Text
- U You
- UFB Ultra Fine business ("very good")
- UR Your or You're (depending on context) Alt: YR
- URS Yours
- VX Voice, phone
- VY Very
- W Watts
- WA Word after
- WB Word before
- WDS Words
- WID With
- WKD Worked
- WKG Working
- WL Will
- WUD Would
- WX Weather
- XCVR Transceiver
- XMTR Transmitter
- XYL Wife (ex-YL) (Extra Young Lady)
- YF Wife
- YL Young lady
- YR Your or You're (depending on context) Alt: UR
- Z Zulu time i.e. UTC (GMT)
- ZX Zero beat
- 33 Used as a greeting between YLs (as half of an 88)
- 44 Hand shake, half of 88. 55 Wishing success (originates from German "Viele Punkte" -- Many dots/points)
- 72 Best Wishes QRP (Low Power) often used by low power station operators (5W or less)
- 73 Best regards
- 77 Long Live CW (Morse Code), wishing you many happy CW contacts
- 88 Love and kisses
- 99 Get lost!

Q Codes

Remember that each of these codes has two meanings: it can be asked as a **question** or given as an **answer**. In the list below its more common form is used. Some of the codes listed here are very rare and you may never hear them used on the Amateur bands as they are/were primarily used in maritime or air navigation (see the link to the ITU).

- QOA - Can you communicate by radiotelegraphy?
- QOT - How many minutes until we can exchange traffic?
- QRA - What is the call sign of your station?
- QRB - How far approximately are you from my station?
- QRD - Where are you bound for and where are you from?
- QRE - What is your estimated time of arrival at ... (place)?
- QRF - Are you returning to ... (place}?
- QRG - My exact frequency is …
- QRH - Your frequency varies.
- QRI - The tone of your transmission is … 1. Good - 2. Variable - 3. bad.
- QRK - What is the intelligibility of my signals?
- QRL - I'm busy, please do not interfere / Is this frequency in use?
- QRM - Interference from other stations.
- QRN - Interference from static or atmospheric noise.
- QRO - Shall I increase transmitter power?
- QRP - Shall I reduce transmitter power? / I am using <5W.
- QRQ - Send faster (wpm).
- QRR - Automatic operation (CW).
- QRS - Send more slowly (wpm).
- QRT - Stop sending / my station is closing down.
- QRU - I have no further messages.
- QRV - I'm ready.
- QRW - Shall I inform ... that you are calling him on ... kHz?
- QRX - Please wait … (minutes). I will call again.
- QRY - Your turn is number …
- QRZ - Who is calling me?
- QSA - Signal strength (RST).
- QSB - Your signal is fading.
- QSD - Is my keying defective?
- QSF - Have you effected a rescue.
- QSG - Send … telegrams at a time.
- QSK - Can you hear me between your signals?
- QSL - I acknowledge receipt.
- QSM - Repeat your last message.
- QSN - I heard you on ... KHz.
- QSO - Can you communicate directly with …. / A 2-way contact.
- QSP - I will relay the message.
- QSQ - Do you have a doctor on board?
- QSR - Repeat your call on the calling frequency; did not hear you.
- QSS - I will use the frequency ... to call.
- QSU - Send or reply on this frequency ... KHz/MHz.
- QSV - Send a series of Vs on this frequency.
- QSX - Will you listen to ...... on ...... Mhz.
- QSY - Change frequency / move to … kHz.
- QSZ - Repeat each word or group twice.
- QTA - Cancel message n°....
- QTC - I have .... messages for you.
- QTE - Your true bearing from me is …
- QTH - My location is ....
- QTI - My true course is (degrees).
- QTJ - My speed is ... knots / km/h.
- QTL – My true heading is ... degrees.
- QTN – I departed from ... (place) at ... hours.
- QTP – I'm going to enter port / land.
- QTQ - Use International Code of Signals.
- QTR - The correct time is …
- QTS - I will send my call sign for … seconds.
- QTU - My station is open from ... to ... hours.
- QTW - The condition of survivors is ...
- QTX - My station will remain on until further notice.
- QTZ - I am continuing the search.
- QUA - Here is news of (call sign).
- QUB - Here is the requested information: ...
- QUC - The n° of the last message received was...
- QUD - I have received the urgency signal sent by...
- QUF - I have received the distress signal sent by...
- QUH - The present barometric pressure at sea level is ...
- QUM - Normal working may be resumed.
- QUS - I have sighted survivors/wreckage at position…
- QUX - I have the following navigational/gale warnings in force...
- QUZ - Distress phase still in force; restricted working may be resumed.

QSO examples
review and discuss with your 'classmates'

It is best to listen to real contacts on the radio but, to start with, here are some examples of what you can hear. Look at the structure, the use of prosigns, abbreviations, etc.

1 *A short QSO.*

CQ CQ CQ DE EC6PG EC6PG EC6PG <AR> <K>

    EC6PG DE EA7HYD EA7HYD <K>

EA7HYD DE EC6PG TNX FER CALL <BT> UR RST 599 599 <BT> HR QTH MALLORCA MALLORCA <BT> NAME XISCO XISCO SO HW CPI? EA7HYD DE EC6PG <KN>

    EC6PG DE EA7HYD TNX FER RPT <BT> SLD CPI UR RST 589 589 <BT> NAME DAVID DAVID <BT> QTH NR CADIZ CADIZ <BT> HW CPI? EC6PG DE EA7HYD <KN>

EA7HYD DE EC6PG FB DAVID <BT> HP CUAGN <BT> VY 73 TO U <AR> EC6PG DE EA7HYD <SK>

    EC6PG DE EA7HYD TNX FER FB QSO XISCO <BT> HPE CUAGN <BT> GL ES GDX <BT> 73 EC6PG DE EA7HYD <SK> TU E E

E E

2 *In this QSO we see how <BK> is used in a fluid conversation between operators who know each other but, nevertheless, from time to time they still send their callsigns.*

CQ CQ CQ DE EA2CW EA2CW EA2CW <AR> <K>

    EA2CW DE EA3NN EA3NN <KN>

EA3NN DE EA2CW GN DR OM DANY = NICE 2 LSTN U AGN = HW UR SON? <BK>

    <BK> OK DR OM MIKEL SON IS FB = UR AMP IS OK ES WILL SEND U NXT WEEK <BK>

<BK> OK DANY TKS FOR GUD NEWS = MNI TKS FER UR WRK = EA3NN DE EA2CW <KN>

    EA2CW DE EA3NN OK ALL = NW PREP TRIP 2 MELILLA = ANTS READY = IC 7300 IS OK = CW KEY ALSO OK = HW? <BK>

<BK> DE EA2CW = OK DANY HR ALSO READY 2 MELILLA = HPE CU THERE = NW GO 2 CLASS = CU SOON = EA3NN DE EA2CW <SK>

    <BK> DE EA3NN = OK MIKEL = GL IN CLASS = CU 73 EA2CW DE EA3NN <AR><SK> E E

E E

3 *A QSO with reception issues. (# = QRM, QRN, QSB...).*

QRL? QRL?
CQ CQ CQ DE W6XYZ W6XYZ W6XYZ <AR> <K>
CQ CQ CQ DE W6XYZ W6XYZ W6XYZ <K>
   W6XYZ DE I#3A## I##AAA <K>
I? RPT RPT PSE DE W6XYZ <K>
   W6XYZ DE IW3AAA IW3AA# ##3AAA <K>
IW3AAA DE W6XYZ GM DR OM ES TNX FER CALL = UR RST IS 359 35N
= MY QTH QTH IS ROMA ROMA = MY NAME NAME IS JOHN JOHN = HW ? =
W6XYZ DE IW3AAA <KN>
   W6XYZ DE IW3AAA R R GM DR OM JOHN = UR R#T IS #99 599 5NN FB
   = HR QT# IS VER###E###A ####NA ES NAME IS ENZO ENZ# ENZO =
   HW ? #6XY# DE ##3AAA <#N>
<BK> PSE RPT QTH QTH <BK>
   <BK> QTH VERONA VERONA ##RONA <BK>
IW3AAA DE W6XYZ FB DR OM ENZO IN VERONA = HR RIG IS TS50 TS50 =
PWR IS ABT 50W 50W = ANT IS VERT VERT = WX WX IS CLOUDY CLOUDY
= TEMP 15C = IW3AAA DE W6XYZ <KN>
   W6XYZ DE IW3AAA R R OK DR OM J#HN FB = HR RIG IS 7300 73TT =
   ANT IS DIPOLE DI##L# = WX FINE = TEMP ABT 28C = TNX FER FB
   QSO = QSL OK OK = 73 GL GB W6XYZ DE IW3AAA <SK>
VE IW3AAA DE W6XYZ = TNX FER NICE QSO = PSE QSL = MY QSL SURE
VIA BURO =   73 73 ES CUAGN CIAO = IW3AAA DE W6XYZ <SK> E E
   W6XYZ DE IW3AAA TNX FER FB QSO = QSL OK OK = 73 GL GB W6XYZ
   DE IW3AAA <SK> TU E E
E E

4 *Example of a SOTA contact (Summits on the Air).*

CQ SOTA DE RU1FGH/P RU1FGH/P <K>
   RU1FGH/P DE EA3GZA EA3GZA <KN>
EA3GZA UR RST 579 57N REF R9U/CE001 <BK>
   <BK> R R RST 599 5NN <BK>
<BK> R R TU 73
   TU EE
EE RU1FGH/P

* *Normally the reference of
the summit is given every 3
or 5 QSOs.*

touch key made by
EA7HOG

by K4TWJ

5 *Calling to a continent, Oceania (OC).*

CQ OC DE LU4OPQ CQ OC OC DE LU4OPQ LU4OPQ <K>
    LU4OPQ DE XX1XYZ XX1XYZ
CQ OC OC DE LU4OPQ CQ OC OC DE LU4OPQ LU4OPQ PSE ONLY OC <K>
    DE XX1XYZ
XX1XYZ PSE QSY UR NOT IN OC = CQ OC DE LU4OPQ CQ OC DE LU4OPQ LU4OPQ OC <K>
    LU4OPQ DE VK3FGH VK3FGH <KN>
VK3FGH DE LU4OPQ TNX FER CALL = UR RST 599 5NN = NAME ALBERTO ALBERTO = QTH NEUQUEN NEUQUEN = HW? <AR> VK3FGH DE LU4OPQ <KN>
    LU4OPQ DE VK3FGH R R DR OM ALBERTO = UR RST 599 5NN = HR NAME JERRY JERRY = QTH NR AUCKLAND AUCKLAND = BTU = LU4OPQ DE VK3FGH <KN>
VK3FGH DE LU4OPQ OK DR OM JERRY = TNX FER RPRT ES NICE QSO = 73 ES GUD DX <AR> <SK> TU
    LU4OPQ DE VK3FGH ALL OK = MNI TKS = CUAGN 73 LU4OPQ DE VK3FGH TU EE
EE

6 *Contest QSO*

CQ TEST EA2CW → M5A
    EA7HYD   M5A 5NN 024
EA7HYD 5NN 023   NR?
?   024 024
EA7HYD 5NN 023   OK 5NN 017
    5NN 087   TU EA2CW
TU EA2CW

LCWO - www.lcwo.net
recommendations on its use

The LCWO website was launched in 2008 by Fabian Kurz, DJ1YFK. It provides a simple and effective way to learn Morse code, it's free and supported by both desktop and mobile phone browsers (with HTML5 or Adobe Flash Player).

You can start practicing with the **Koch Method Lessons** or with **Code Groups**. The first is perfect to learn a little at a time, at your own pace, and the second is a great way to repeat characters that give you problems so you can build familiarity. Of course, once we have learned the code we can also practice callsigns, words, texts or QTC (for contests like WAE).

Although clearly explained on the website itself, I have noticed that some people encounter difficulties when setting it up. Once you register on the website (name / password), these are my recommendations for the "**Change CW settings**" section. Of course, feel free to set your own preferences:

· Start with **a speed of** 20 wpm, never less (see p.25).

· Set the **effective speed** to 10 wpm or 12 wpm. If you can manage higher then so much the better.
As the lessons progress you can increase this to 20 wpm.
Once you have completed the lessons, you can practice increasing the real and effective speed until you reach ... 35 wpm or more (HI).

· You can set the **tone** (Hz) to a value that you find comfortable to listen to but it's better to leave it set to "random".

· HTML5 is a recommended as the CW player.

· To start with, avoid using the **Transmission prefix / suffix** option (VVV = / AR). If you want to use it, do so after a few lessons otherwise it may confuse you a little.

· Add a few seconds to the **start delay** so that you have a little time after you "start" a lesson to prepare to write.

· The **group length** (number of characters in a group) is 5 by default, there is no need to change it.

· In the right hand column you can select the **characters** you wish to practice when you select "Custom characters" within the Code Groups training. This option is perfect to separately review and practice any numbers, letters or prosigns that you are struggling to learn.

· Remember to **press the Submit button** every time you make a change on this page otherwise your preferences will not be saved.

· Check out the instructions and tips on the web.

Learn CW online, in your web browser!
Koch Method Morse Course, Speed Training,
Text to CW conversion, Statistics, Forum

CW4U.org - 87

Portable, QRP, SOTA

CW is a great choice for portable operating
by Michael Sansom, G0POT

Operating Portable

Whether you are operating from the beach, a park, a mountain top, a car or a canoe, **CW is a great choice for portable operating**. Transceivers can be small, simple and light and QRP CW packs far more punch than QRP SSB on busy bands. Being able to run lower power means smaller batteries, saving space and a lot of weight. Headphones save weight and space.

Equipment

There are a huge range of small CW transceivers including commercial radios, commercial kits and homebrew designs. Classics like the HB-1B and Mountain Topper offer multiple bands, good filtering and a few watts of RF. Commercial kits like the PFR-3 and QCX mean that you can be familiar with the insides of your radio in case you have issues and homebrew allows you to customize the radio to a specific task or your own requirements.

The simpler your station the less that can go wrong so there is a great advantage to using pre tuned antennas for portable operation removing the need for an ATU. However, you never know where you might have to erect an aerial so a small aerial matching unit can prove useful.

Wire dipoles are light and hard to beat but you will need to include feeder. Weight can be reduced by using an End Fed Half Wave (EFHW) wire antenna as a sloper or inverted V as this type of aerial needs just a few feet of coax.

Portable equipment: 10m travel mast, QRP transceiver, wire antennas 20 ,30, 40m, headphones , Lipo batteries...

The Morse key in your shack is a precision instrument and won't take kindly to being transported in a rucksack and being used outdoors in mud & rain or sun & sand. Temperature changes can affect the contact spacing of a key causing the contacts to short out. If you are operating in muddy or sandy locations the ingress of dirt into any part of the key can cause problems.

EA5IW/P

Simple, enclosed keys are a good choice; either straight keys or paddles are fine. Keys like the Te-Ne-Ke can be held in one hand while being operated with the other, have no real moving parts and are simple to adjust in the field.

Portable operators tend to use slightly larger contact spacing than normal on their keys to overcome temperature changes, resting the key on uneven surfaces and trying to send with cold hands!

Summits on the Air – taking Morse code to new heights!

Summits on the Air (SOTA) combines walking, hill/mountain climbing and operating radio...a wonderful combination. Best of all, when you operate SOTA you will feel like rare DX with all the 'chasers' wanting to work you.

As a SOTA Operator you will need to learn to manage a pile up, however, you don't need to worry about 'working split' as SOTA chasers are normally quite well behaved. When a lot of stations call at once simply try to copy a few letters from one of the calls and then send what you got with a question mark like 'AA5?' or 'HYD?'. That should reduce the number of calls that come back to you. Operating is typically contest-like with short exchanges which makes SOTA and contesting a great way to get started making contacts with Morse code. However, at the end of the day, YOU are the 'DX' so YOU are in control. If you want to slow things down or include additional information like names and/or locations in the QSO then that is up to you.

The other advantage of using CW for SOTA is that you can register a planned activation at *sotawatch.org* and if the *Reverse Beacon Network* (RBN) hears you calling "CQ CQ CQ SOTA DE G0POT/P" it can automatically 'spot' you, posting a notification that you are active on the sotawatch website.

The code you'll need to know:

Along with the <BK> prosign you'll need to be familiar with how to send a forward slash '/' as it's used a lot. You'll send /P to indicate you are portable and again when you send a summit reference, like W7I/IC002 (see page 85, example 4).

Simple, enclosed keys are a good choice

Plan IARU
Regions 1 and 2, HF Band
Frequencies and exclusive segments for CW

This plan for the use of the bands for all radio amateurs is a recommendation of the IARU (International Amateur Radio Union), effective as of 2016. Its purpose is to organize in the best way the efficient use of the bands.

CW: Telegraphy (**A1A**) QSO. You can transmit in CW on ALL OF THE BANDS except in the frequencies dedicated to beacons, etc.

Remember that frequencies have no owner and that a recommended frequency is not exclusive.

Only four bands are represented. For more information, see the full plan on the IARU website: *https://www.iaru-r1.org* (r1 = r2 or r3).

REGION 1.-

Band Frequencies (kHz)	Bandwidth (Hz)	Mode	Applications and observations
20 m 14.000 - 14.070 KHz			
14000 - 14060	200	CW	QRS 14055 kHz Centre of Activity
14060 - 14070	200	CW, DM*	QRP 14060 kHz Centre of Activity
30 m 10.100 - 10.130 KHz			
10100 - 10130	200	CW	QRP 10116 kHz / QRS 10125 kHz
40 m 7.000 - 7.040 KHz			
7000 - 7040	200	CW	QRP 7030 kHz / QRS 7035 kHz
80 m 3.500 - 3.580 KHz			
3500 - 3510	200	CW	Priority intercontinental operation
3510 - 3560	200	CW	Contest preferred QRS Center of Activity 3555 kHz
3560 - 3570	200	CW	QRP Center of Activity 3560 kHz
3570 - 3580	200	CW, DM	

*DM (Digital Modes)

REGION 2 .-

Band Frequencies (kHz)	Bandwidth (Hz)	Mode	Applications and observations
20 m 14.000 - 14.101 KHz			
14000-14025	200	CW	Priority intercontinental operation (DX window)
14025-14060	200	CW	CW Contests preferred QRS Center of Activity 14055 kHz
14060-14070	200	CW	QRP Center of Activity 14060 kHz
14070-14089	500	CW,DM	
14089-14099	500	CW,DM	ACDS (Automatic Controlled Data Stations)
14099-14101	200	CW	IBP (exclusive, International Beacon Project)
30 m 10.100 - 10.150 KHz			
10100 - 10130	200	CW	QRP 10116 kHz / QRS 10125 kHz
10130 - 10140	500	CW	ACDS
10140 - 10150	2700	CW, DM	
40 m 7.000 - 7.050 KHz			
7000-7025	200	CW	Priority intercontinental operation (DX window)
7025-7040	200	CW	QRP 7030 kHz / QRS 7035 kHz
7040-7047	500	CW, DM	
7047-7050	500	CW, DM	ACDS
80 m 3.500 - 3.600 KHz			
3500-3510	200	CW	Priority intercontinental operation (DX window)
3510-3560	200	CW	QRS Center of Activity 3555 kHz CW contest preferred
3560-3570	200	CW	QRP Center of Activity 3560 kHz
3570-3580	200	CW	
3580-3590	500	CW, DM	
3590-3600	500	CW, DM	ACDS

danish key

Links, videos...
you will find additional information on the internet

Online methods:
- Learn CW online
  http://lcwo.net/

Instructional videos:
- US Navy Training Video - Technique Of Hand Sending Morse Code.
  https://www.youtube.com/watch?v=NCST3kW4VOo
- International Morse Code Hand Sending 1966 US Army Training Film
  https://www.youtube.com/watch?v=e26XQZNVfs8
- Morse Code Music... The Rhythm of the Code
  https://www.youtube.com/watch?v=0OvVru8MHdI
- Iambic sending demostration
  https://youtu.be/r7RWgcDLceQ

Information about Morse code:
- IARU:
  http://www.iaru.org
- International Comunication Union ITU - Q Codes
  https://www.itu.int/dms_pubrec/itu-r/rec/m/R-REC-M.1172-0-199510-I!!PDF-S.pdf
- Straight Key Century Club (SKCC)
  http://www.skccgroup.com/member_services/beginners_corner/
- Learning Morse Code - ARRL
  http://www.arrl.org/learning-morse-code

Free programs:
- G4FON Morse Trainer
  http://www.g4fon.net/CW%20Trainer.htm
- Just Learn Morse Code
  http://www.justlearnmorsecode.com/
- Morse Runner
  http://www.dxatlas.com/MorseRunner/
- RufzXP - practice with callsings
  http://www.rufzxp.net/
- ICWOIP - QSO online with other QRQcw ops
  https://sites.google.com/site/icwoip

SDR online receivers:
- List
  http://websdr.org/

J38 key

Bibliography

there are many texts about the morse code, here are a few recommendations

- The art and skill of radio telegraphy
  William G. Pierpont, N0HFF
  http://www.tasrt.ca/bookdown.html

- How to make a morse code contact
  Chris R. Burger, ZS6EZ
  http://zs6ez.org.za/tutorial/cw-qso.pdf

- Morse Code Practice for Radio Amateurs
  D. A. Campbell, N1CWR
  http://www.orarc.net/Morse.pdf

- ethics and operating procedures for the radio amateur
  Mark Demeuleneere, ON4WW
  http://www.hamradio-operating-ethics.org/versions

Download a free QSO template poster with the main codes, abbreviations, MP3 files and more at the address:
www.cw4u.org

See Facebook webpage:
www.facebook.com/cw4dummies

Send you opinions, suggestions or comments:
cw4u @ hotmail.com

On the web www.cw4u.org
You have the updated link list..

unknown

QRA: .... DATE: ..../......./....... QTH: ....
MODE: .... RIG: .... PWR: ....W ANT: ....

TIME	FREQ.	STATION WORKED	REPORT SENT	REPORT RECEIVED	COMMENTS

QRA: ....
DATE: ...... / ...... / ......
QTH: ....

MODE: ....
RIG: ....
PWR: ...... W
ANT: ....

TIME	FREQ.	STATION WORKED	REPORT SENT	REPORT RECEIVED	COMMENTS

> **THERE ARE TWO MAIN STEPS TO LEARN MORSE CODE:**
> **BE TENACIOUS AND LISTEN** ⚠️

I am positive that if you are determined and follow the instructions in this book, you will soon be able to start sending and receiving in Morse code. Let's go...you can do it.